The Universe Within

NEIL SHUBIN

The Universe Within

A Scientific Adventure

ALLEN LANE
an imprint of
PENGUIN BOOKS

ALLEN LANE

Published by the Penguin Group
Penguin Books Ltd, 80 Strand, London WC2R 0RL, England
Penguin Group (USA) Inc., 375 Hudson Street, New York, New York 10014, USA
Penguin Group (Canada), 90 Eglinton Avenue East, Suite 700, Toronto, Ontario, Canada M4P 2Y3
(a division of Pearson Penguin Canada Inc.)
Penguin Ireland, 25 St Stephen's Green, Dublin 2, Ireland (a division of Penguin Books Ltd)
Penguin Group (Australia), 707 Collins Street, Melbourne, Victoria 3008, Australia
(a division of Pearson Australia Group Pty Ltd)
Penguin Books India Pvt Ltd, 11 Community Centre, Panchsheel Park, New Delhi – 110 017, India
Penguin Group (NZ), 67 Apollo Drive, Rosedale, Auckland 0632, New Zealand
(a division of Pearson New Zealand Ltd)
Penguin Books (South Africa) (Pty) Ltd, Block D, Rosebank Office Park,
181 Jan Smuts Avenue, Parktown North, Gauteng 2193, South Africa

Penguin Books Ltd, Registered Offices: 80 Strand, London WC2R 0RL, England

www.penguin.com

First published in the United States by Pantheon Books,
a division of Random House, Inc. 2013
First published in Great Britain by Allen Lane 2013
001

Copyright © Neil Shubin, 2013

The moral right of the author has been asserted

Printed in Great Britain by Clays Ltd, St Ives plc

A CIP catalogue record for this book is available from the British Library

ISBN: 978–1–846–14220–8

FOR MICHELE, NATHANIEL, AND HANNAH

CONTENTS

PROLOGUE

Having spent the better part of my working life staring at rocks
on the ground, I've gained a certain perspective on life and the
universe. My professional aspiration—uncovering clues to the
making of our bodies—lies inside the baked desert floor or deep
within the frozen Arctic. While this ambition may seem eccentric, it is not much different from that of colleagues who peer
at the light of distant stars and galaxies, map the bottom of the
oceans, or chart the surface of barren planets in our solar system.
What weaves our work together are some of the most powerful ideas that mankind has ever developed, ones that can explain
how we and our world came to be.

These notions inspired my first book, *Your Inner Fish*. Inside
every organ, cell, and piece of DNA in our bodies lie over
3.5 billion years of the history of life. Accordingly, clues to the
human story reside within impressions of worms in rock, the
DNA of fish, and clumps of algae in a pond.

While I was thinking about that book, it became clear
that worms, fish, and algae are but gateways to ever deeper
connections—ones that extend back billions of years before the
presence of life and of Earth itself. Written inside us is the birth
of the stars, the movement of heavenly bodies across the sky,
even the origin of days themselves.

During the past 13.7 billion years (or so), the universe came
about in the big bang, stars have formed and died, and our planet

congealed from matter in space. In the eons since, Earth has circled the sun while mountains, seas, and whole continents have come and gone.

Discovery after discovery in the past century has confirmed the multibillion-year age of Earth, the sheer vastness of the cosmos, and our species' humble position in the tree of life on our planet. Against this backdrop, you could legitimately wonder if it is part of the job description of scientists to make people feel utterly puny and insignificant in the face of the enormity of space and time.

But by smashing the smallest atoms and surveying the largest galaxies, exploring rocks on the highest mountains and in the deepest seas, and coming to terms with the DNA inside every species alive today, we uncover a sublimely beautiful truth. Within each of us lie some of the most profound stories of all.

THE UNIVERSE WITHIN

CHAPTER ONE

ROCKING OUR WORLD

Viewed from the sky, my companion and I must have looked like two black specks perched high on a vast plain of rock, snow, and ice. It was the end of a long trek, and we were slogging our way back to camp on a ridge sandwiched between two of the greatest ice sheets on the planet. The clear northern sky opened a panorama that swept from the pack ice of the Arctic Ocean in the east to the seemingly boundless Greenland ice cap to our west. After a productive day prospecting for fossils and an exhilarating hike, and with the majestic vista around us, we felt as if we were walking on top of the world.

Our reverie was abruptly cut short by a change in the rocks beneath our feet. As we traversed the bedrock, brown sandstones gave way to ledges of pink limestone that, from our earlier discoveries, became an auspicious sign that fossils were in the neighborhood. After we spent a few minutes peering at boulders, alarm bells went off; my attention was pulled to an unusual glimmer flashing from a corner of a melon-sized rock. Experience in the field taught me to respect the sensation triggered by these moments. We had traveled to Greenland to hunt for small fossils, so I hunched over my magnifying lens to scan the rock closely. The sparkle that arrested me sprang from a little white spot, no bigger than a sesame seed. I spent the better part of five minutes curled up with the rock close to my eyes before passing it to my colleague Farish for his expert opinion.

Concentrating attention on the fleck with his lens, Farish froze solid. His eyes shot back to me with a look of pent-up emotion, disbelief, and surprise. Rising from his crouch, he took off his gloves and launched them about twenty feet in the air. Then he nearly crushed me with one of the most titanic bear hugs I have ever received.

Farish's exuberance made me forget the near absurdity of feeling excitement at finding a tooth not much bigger than a grain of sand. We found what we had spent three years, countless dollars, and many sprained ligaments looking for: a 200-million-year-old link between reptiles and mammals. But this project was no miniature trophy hunt. The little tooth represents one of our own links to worlds long gone. Hidden inside these Greenlandic rocks lie our deep ties to the forces that shaped our bodies, the planet, even the entire universe.

Seeing our connections to the natural world is like detecting the pattern hidden inside an optical illusion. We encounter bodies, rocks, and stars every day of our lives. Train the eye, and these familiar entities give way to deeper realities. When you learn to view the world through this lens, bodies and stars become windows to a past that was vast almost beyond comprehension, occasionally catastrophic, and always shared among living things and the universe that fostered them.

How does such a big world lie inside this tiny tooth, let alone inside our bodies? The story starts with how we ended up on that frozen Greenlandic ridge in the first place.

Imagine arriving at a vista that extends as far as the eye can see, knowing you are looking inside it for a fossil the size of the period that ends this sentence. If fossil bones can be small, so too are whole vistas relative to the surface area of Earth. Knowing how to find past life means learning to see rocks not as static objects but as entities with a dynamic and often violent history. It also means understanding that our bodies, as well as our entire world, represent just moments in time.

The playbook that fossil hunters use to develop new places to look has been pretty much unchanged for the past 150 years. Intellectually, it is as simple as it gets: find places on the planet that have rocks of the right age to answer whatever question interests you, rocks of the type likely to hold fossils, and rocks exposed on the surface. The less you have to dig, the better. This approach, which I described in *Your Inner Fish,* led me and my colleagues, in 2004, to find a fish at the cusp of the transition to life on land.

As a student in the early 1980s, I gravitated to a team that had developed tools to make headway finding new places to hunt fossils. Their goal was to uncover the earliest relatives of mammals in the fossil record. The group had found small shrewlike fossils and their reptilian cousins in a number of places in the American West, but by the mid-1980s their success had brought them to an impasse. The problem is best captured by the jest "Each newly discovered missing link creates two new gaps in the fossil record." They had done their share of creating gaps and were now left with one in rocks about 200 million years old.

The search for fossil sites is aided by economics and politics. With the potential for significant oil, gas, and mineral discoveries, there are incentives for countries to catalog and map the geology exposed inside their borders. Consequently, virtually any geological library holds journal articles, reports, and, one hopes, maps detailing the age, structure, and mineral content of the rocks exposed on the surface of different regions. The challenge is to find the right maps.

Professor Farish A. Jenkins Jr. led the team at Harvard's Museum of Comparative Zoology. Fossil discovery was the coin of the realm for him and his crew, and it started in the library.

Farish's laboratory colleagues Chuck Schaff and Bill Amaral were key in this effort; they had honed their understanding of geology to predict likely places to make discoveries, and, importantly, they trained their eyes to find really small fossils. Their relationship often took the form of a long, friendly argument: one would propose a new idea while the other would relentlessly try to quash it. If the idea held up under their largely amiable tit for tat, then they would both line up behind the proposal and take it to Farish, with his keen logistic and scientific sense, for vetting.

One day in 1986, while chewing the fat with Chuck, Bill found a copy of the *Shell Oil Guide to the Permian and Triassic of the World* on Chuck's desk. Paging through the volume, Bill spotted a map of Greenland, with a little hatched area of Triassic rocks on the eastern coast at a latitude of about 72 degrees north, roughly that of the northernmost tip of Alaska. Bill kicked things off by proclaiming that this could be a prime next area to work. The usual argument ensued, with Chuck denying that the rocks were the right type, Bill responding, and Chuck countering.

By dumb luck, Chuck had the means to end the debate right on his bookshelf. A few weeks earlier, he was trolling through the library discards and pulled out a paper titled "Revision of Triassic Stratigraphy of the Scoresby Land and Jameson Land Region, East Greenland," authored by a team of Danish geologists in the 1970s. Little did anyone know at the time, but this freebie, saved from the trash heap, was to loom large in our lives for the next ten years. Virtually from the minute Bill and Chuck looked at the maps in the reprint, the debate was over.

My graduate student office was down the hall, and as was typical for that time in the late afternoon, I swung by Chuck's office to see what was what. Bill was hovering about, and it was clear that some residue from one of their debates remained in the air. Bill didn't say much; he just slapped Chuck's geological reprint down in front of me. In it was a map that showed exactly what we had hoped for. Exposed on the eastern coast of Greenland, across the ocean from Iceland, were the perfect kinds of rocks in which to find early mammals, dinosaurs, and other scientific goodies.

The maps looked exotic, even ominous. The east coast of Greenland is remote and mountainous. And the names evoke explorers of the past: Jameson Land, Scoresby Land, and Wegener Halvø. It didn't help matters that I knew that a number of explorers had perished during their trips there.

Fortunately, the expeditions that transpired ultimately rested on Farish's, Bill's, and Chuck's shoulders. With about sixty years of fieldwork between them, they had developed a deep reservoir of hard-earned knowledge about working in different kinds of field conditions. Of course, few experiences could have prepared us for this one. As a famed expedition leader once told me, "There is nothing like your first trip to the Arctic."

I learned plenty of lessons that first year in Greenland, ones that were to become useful when I began running my own Arctic expeditions eleven years later. By bringing leaky leather boots, a small used tent, and a huge flashlight to the land of mud, ice, and the midnight sun, I made so many bad choices that first year that I remained smiling only by reciting my own motto, "Never do anything for the first time."

The most nerve-racking moment of that inaugural trip came when selecting the initial base camp, a decision made in a fleeting moment while flying in a helicopter. As the rotors turn, money flies out the window, because the costs of Arctic helicopters can be as high as three thousand dollars per hour. On a paleontology budget, geared more to beat-up pickups than to Bell 212 Twin Hueys, that means wasting no time. Once over a promising site revealed by the maps back in the laboratory, we rapidly check off a number of important properties before setting down. We need to find a patch of ground that is dry and flat yet still close to water for our daily camp needs, far enough inland so that polar bears aren't a problem, shielded from the wind, and near exposures of rock to study.

The Greenland crew clockwise from top left: Farish, military trim; Chuck, wise fossil finder; Bill, man who makes things happen in the field; and me. I made a lot of bad choices that first year (note hat).

We had a good idea of the general area from the maps and aerial photographs, and ended up setting down on a beautiful little patch of tundra in the middle of a wide valley. There were creeks from which we could draw water. The place was flat and dry, so we could pitch our tents securely. It even had a gorgeous view of a snowy mountain range and glacier on the eastern end of the valley. But we would soon discover a major shortcoming. There were no decent rocks within easy walking distance.

Once camp was established to our satisfaction, we set off each day with one goal in mind: to find the rocks. We'd climb the highest elevations near camp and scan the distance with binoculars for any of the exposures that figured so prominently in the paper Bill and Chuck had found. Our search was eased by the fact that the rock layers were collectively known as red beds for their characteristic hue.

With red rock on our minds, we went off in teams, Chuck and Farish climbing hills to give them views of the southern rocks, Bill and I setting off for places that would reveal those to the north. Three days into the hunt, both teams returned with the same news. Out in the distance, about six miles away to the northeast, was a sliver of red. We'd argue about this little outcrop of rock, scoping it with our binoculars at every opportunity for the remainder of the week. Some days, when the light was right, it seemed to be a series of ridges ideal for fossil work.

It was decided that Bill and I would scout a trail to get to the rocks. Since I didn't know how to walk in the Arctic, and had made an unfortunate boot selection, the trek turned out to be an ordeal—first through boulder fields, then across small glaciers, and pretty much through mud for the rest of the way. The mud formed from wet clay that made an indelicate *glurp* as we extricated our feet from each step. No footprint remained, only a jiggling viscous mass.

In three days of testing routes, we plotted a viable course to the promising rocks. After a four-hour hike, the red sliver in our binocular view from camp turned out to be a series of cliffs, ridges, and hillocks of the exact kind of rock we needed. With any luck, bones would be weathering out of the rock's surface.

The goal now became to return with Farish and Chuck, doing the hike as fast as possible to leave enough time to hunt for bones before having to turn back home. Arriving with the whole crew, Bill and I felt like proud homeowners showing off our property. Farish and Chuck, tired from the hike but excited about the prospect of finding fossils, were in no mood to chat. They swiftly got into the paleontological rhythm of walking the rocks at a slow pace, eyes on the ground, methodically scanning for bone at the surface.

Bill and I set off for a ridge about half a mile away that would give us a view of what awaited us even farther north. After a small break, Bill started to scan the landscape for anything of interest: our colleagues, polar bears, other wildlife. He stopped scanning and said, "Chuck's down." Training my binoculars on his object, I could see Chuck was indeed on his hands and knees methodically crawling on the rock. To a paleontologist this meant one thing: Chuck was picking up fossil bones.

Our short amble to Chuck confirmed the promise of the binocular scan; he had indeed found a small piece of bone. But our hike to this little spot had taken four hours, and we now had to head back. We set off, with Farish, Bill, Chuck, and me in a line about thirty feet apart. After about a quarter of a mile something on the ground caught my eye. It had a sheen that I'd seen before. Dropping to my knees like Chuck an hour earlier, I saw it in its full glory, a hunk of bone the size of my fist. To the left was more bone, to the right even more. I called to Farish, Bill, and Chuck. No response. Looking up, I knew why. They were also on their hands and knees. We were all crawling in the same colossal field of broken bones.

At summer's end, we returned boxes of these fossil bones to the lab, where Bill put them together like a three-dimensional jigsaw puzzle. The creature was about twenty feet long, with a series of flat leaf-shaped teeth, a long neck, and a small head. The beast had the diagnostic limb anatomy of a dinosaur, albeit a relatively small one.

This kind of dinosaur, known as a prosauropod, holds an

important place in North American paleontology. Dinosaurs in eastern North America were originally discovered along streams, railroad lines, and roads, the only places with decent exposures of rocks. The eminent Yale paleontologist Richard Swann Lull (1867–1957) found a prosauropod in a rock quarry in Manchester, Connecticut. The only problem was that it was the back end. The block containing the front end, he was chagrined to learn, had earlier been incorporated into the abutment of a bridge in the town of South Manchester. Undeterred, Lull described the dinosaur from its rear end only. When the bridge was demolished in 1969, the other fragments came to light. Who knows what fossil dinosaurs remain to be discovered deep inside Manhattan? The island's famous brownstone town houses are made of this same kind of sandstone.

The hills in Greenland form large staircases of rock that not only break boots but also tell the story of the stones' origins. Hard layers of sandstones, almost as resistant as concrete, poke out from softer ones that weather away more quickly. Virtually identical staircases lie farther south; matching sandstones, siltstones, and shales extend from North Carolina to Connecticut all the way to Greenland. These layers have a distinctive signature of faults and sediment. They speak of places where lakes sat inside steep valleys that formed as the earth fractured apart. The pattern of ancient faults, volcanoes, and lake beds in these rocks is almost identical to the great rift lakes in Africa today—Lake Victoria and Lake Malawi—where movements inside Earth cause the surface to split and separate, leaving a gaping basin filled by the water of lakes and streams. In the past, rifts like these extended all the way up the coast of North America.

From the beginning, our whole plan was to follow the trail of the rifts. Knowing that the rocks in eastern North America contained dinosaurs and small mammal-like creatures gave us the aha moment with Chuck's geological reprint. That, in turn, led us north to Greenland. Then, once in Greenland, we pursued

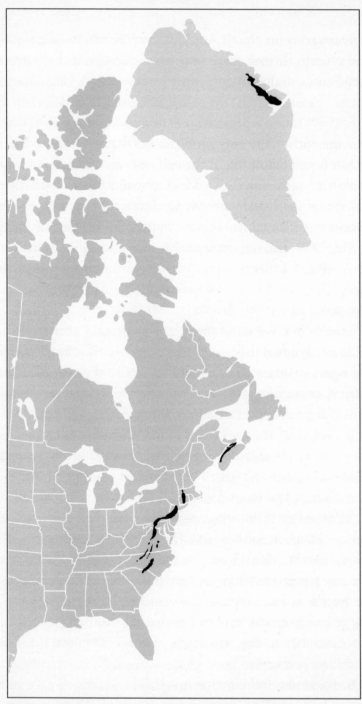
Follow matching rocks (black) to find fossils. Success working rocks in Connecticut and Nova Scotia led us to Greenland.

the discoveries on the ground like pigeons following a trail of bread crumbs. It took three years, but clues in the red beds ultimately led us to that frozen ridge I trekked with Farish.

From the top of the ridge, the tents of camp looked like tiny white dots just below the horizon. The crest was windswept, but the bluff of pink limestone on which we were seated formed a quiet shield for Farish and me to assess the discovery. Farish's jubilation confirmed my hunch that the white spot in the rock was indeed a mammal tooth. With the characteristic pattern of three cusps and two roots, it was a dead ringer for one of these beasts.

Armed with confidence that came from this first discovery, the team looked widely across east Greenland, eventually finding better mammal fossils in subsequent years. The fossils came from a small shrewlike animal about half the size of a house mouse. Although it lacks the sort of awe-inspiring skeleton that would grace a museum rotunda, its beauty lies elsewhere.

This is one of the first creatures in the fossil record with our kind of teeth: those with cutting surfaces defined by cusps that occlude on upper and lower teeth with a tooth row subdivided into incisors, canines, and molars. It has an ear that is like ours also, containing little bones that connect the eardrum to the inner ear. Its skull pattern, shoulder, and limb are also decidedly mammalian. We don't know for sure, but it likely had hair and other mammalian features such as milk-producing glands. Every time we chew, hear high pitches, or rotate our hands, we use parts of our anatomy that can be traced through primates and other mammals to the structures in these little creatures from 200 million years ago.

The rocks also tie us to the past; rifts in Earth, like those that led us to find fossil mammals in Greenland, have left their traces in our bodies as much as they have in the crust of the planet. The

Greenlandic rocks are like one page in a vast library of volumes that contain the story of our world. Billions of years of history preceded that little tooth, and 200 million years have followed it. Through eons on Earth, seas have opened and closed, mountains have risen and eroded, and asteroids have come crashing down as the planet has coursed its way through the solar system. The layers of rock record era after era of changes to the climate, atmosphere, and crust of the planet itself. Transformation is the order of the day for the world: bodies grow and die, species emerge and go extinct, while every feature of our planetary and celestial home undergoes gradual change or episodes of catastrophic revolution.

Rocks and bodies are kinds of time capsules that carry the signature of great events that shaped them. The molecules that compose our bodies arose in stellar events in the distant origin of the solar system. Changes to Earth's atmosphere sculpted our cells and entire metabolic machinery. Pulses of mountain building, changes in orbits of the planet, and revolutions within Earth itself have had an impact on our bodies, minds, and the way we perceive the world around us.

Just like human bodies, this book is organized as a time line. We begin our story about 13.7 billion years ago, when the universe emerged in the big bang. Then we follow the history of our small corner of it to look at the consequences of the formation of the solar system, the moon, and the globe of Earth on the organs, cells, and genes inside each of us.

CHAPTER TWO

BLASTS FROM THE PAST

$H_{375,000,000}$ $O_{132,000,000}$ $C_{85,700,000}$ $N_{6,430,000}$ $Ca_{1,500,000}$
$P_{1,020,000}$ $S_{206,000}$ $Na_{183,000}$ $K_{177,000}$ $Cl_{127,000}$ $Mg_{40,000}$ $Si_{38,600}$
$Fe_{2,680}$ $Zn_{2,110}$ Cu_{76} I_{14} Mn_{13} F_{13} Cr_7 Se_4 Mo_3 Co_1

is a formula of elements that make up the human body. We are a very select mix of atoms. Bodies are mostly hydrogen: for every atom of cobalt, for example, there are almost 400 million of hydrogen. By weight, we contain such a large amount of oxygen and carbon that we are virtually unique in the known universe.

One particular element missing from bodies tells a big story. Helium, the second-most abundant atom in the entire universe, has an internal structure that leaves it no room to trade electrons with others. Unable to make these exchanges, it cannot participate in the chemical reactions that define life—metabolism, reproduction, and growth. On the other hand, oxygen and carbon are about twenty times rarer than helium. But unlike helium, these atoms can easily interact with different elements to form the variety of chemical bonds that are essential in living matter. Reactivity is the order of the day for the common atoms of bodies. Loners need not apply.

The relative proportion of atoms is only a part of what defines our bodily structure. Bodies are organized like a set of Russian nesting dolls: tiny particles make atoms, groups of atoms make molecules, and molecules assemble and interact in different ways to compose our cells, tissues, and organs. Each level of organiza-

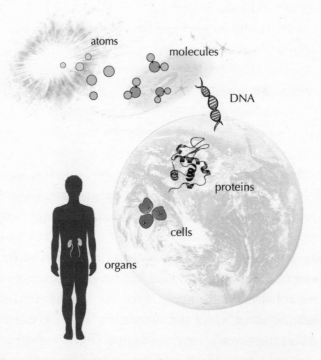

The Russian nesting dolls of all matter: tiny particles make atoms, atoms make molecules, and molecules make ever-larger entities.

tion brings new properties that are greater than the sum of its parts. You could know everything about each of the atoms inside your own liver, but that will not tell you how a liver works. Hierarchical architecture, smaller things making larger entities with new defining properties, is the basic way our world is organized and ultimately reveals our deepest connections to the universe, solar system, and planet.

Pick up a scientific journal in biology nowadays, and you stand a good chance of seeing a tree of relatedness. Every creature, from human to Thoroughbred to prize Hereford, has a pedigree—its family tree. These trees define how closely related living things are: first cousins are more closely related to one another than they are to second cousins. Knowing the pedigree becomes the basis for understanding how different creatures are connected to one another, how species came about, even why certain individuals may be more susceptible to disease than others. This is why doctors take family histories in medical exams.

A critical insight of modern biology is that our family history extends to all other living things. Unlocking this relationship means comparing different species with one another in a very precise way. An order to life is revealed in the features creatures have: closely related ones share more features with each other than do those more distantly related. A cow shares more organs and genes with people than it does with a fly: hair, warm-bloodedness, and mammary glands are shared by mammals and absent in insects. Until somebody finds a hairy fly with breasts, we would consider flies distant relatives to cows and people. Add a fish to this comparison, and we discover that fish are more closely related to cows and people than they are to flies. The reason is that fish, like people, have backbones, skulls, and appendages, all of which are lacking in flies. We can follow this logic to add species after species and find the family tree that relates people, fish, and flies to the millions of other species on the planet.

But why stop at living things?

The sun burns hydrogen. Other stars burn oxygen and carbon. The fundamental atoms that make our hands, feet, and brains serve as the fuel for stars. It isn't merely the atoms in our bodies that extend across the far reaches of the universe: molecules that make our bodies are found in space. The building blocks for the proteins and larger molecules that make us—amino acids and nitrates—rain down to Earth in meteorites and lie on the rocky crust of Mars or on the moons of Jupiter. If our chemical cousins are in the stars, meteors, and other heavenly bodies, then clues to our deepest connections to the universe must lie in the sky above our heads.

Detecting patterns in the sky—the shapes of galaxies, the features on planets, or the components of a binary star—is no easy task. Eyes take some time to adjust to the dark, but so too does perception. You need to train the eye to perceive faint patterns in the night sky. When it comes to deciphering fuzzy patches of stars through a telescope or binoculars, imagination and expectation have a way of conjuring mirages in the void. Removing these and actually seeing dim objects in space means emphasizing peripheral vision, the most sensitive light-gathering part of our eyes, to pick up faint light and discriminate fuzzy patches from one another. As we learn to see the sky, color, depth, and shape emerge in the world above our heads much like when a fossil bone pops into view on a dusty desert floor beneath our feet.

Discriminating celestial objects is merely the first step in learning to see the sky. Like a painting that has graced a house for generations, the stellar landscape we encounter today is much the same as that witnessed by our parents, grandparents, even our apelike ancestors. Generations of humans have not only seen the sky but, over time, built new ways of perceiving our connection to it.

———

Our relationship to the stars changed dramatically because of breakthroughs made by the Harvard Computers at the turn of the twentieth century. Edward Charles Pickering, then director of the Harvard College Observatory, had a problem that required serious computation and analysis. The observatory was collecting reams of pictures of constellations, stars, and nebulae—so many that just managing and plotting the images was a daunting task. Of course, digital computers as we know them didn't exist at this time and the calculations had to be done by hand. Pickering was famously cheap and once declared in a fit of exasperation with his existing staff that he could hire his maid to do this work at half the cost. He fell in love with his new idea and ended up pressing his real maid, Williamina Fleming, into service at the observatory.

At age twenty-one and with a young son, Williamina Fleming was abandoned by her husband, leaving her penniless and without a trade. Pickering first hired her to clean house. Then, after his boast, he brought her to the observatory to manage his celestial images. Upon receipt of a large donation, Pickering was able to add a number of other women to the group. What Pickering could never have planned was that from this team grew some of the greatest astronomers of the time, or any time for that matter. These women collectively became known as the Harvard Computers: they sat with the raw data of astronomy, pictures of the heavens, and made sense of them.

Henrietta Leavitt, the daughter of a Congregational minister, came to the observatory in 1895, first volunteering and later earning a salary of thirty cents an hour. She developed a love for astronomy in school, a passion that served her well during the long years she had the mind-numbing task of cataloging photographic plate after plate of stars and nebulae.

As Leavitt knew, the different stars in the sky vary in color and magnitude of their light. Some stars are dim or small, others bright and big. Of course, there was no real way of knowing

Edward Charles Pickering (upper row) and the "Harvard Computers." Williamina Fleming is in the front row, third from left; Henrietta Leavitt is just to the right of Pickering.

what magnitude meant for the real brilliance of a star, because an
apparently dim star could be a big and bright one far away or a
faint one relatively close.

Leavitt became fascinated by one type of star that changed
regularly from bright to dim over the course of days or months.
Mapping seventeen hundred stars, she charted every property
she could measure: how bright they were, where they sat in the
sky, and how rapidly these variable stars went from bright to dim.
With all of these data, Leavitt uncovered an important regularity: there is a constant relationship between how fast some stars
cycle from bright to dim and their *real brightness*.

Leavitt's idea seems awfully esoteric, but it is profound.
Starting with the principle that light travels at a constant speed,
and knowing how bright the star actually was and how bright it
appeared, meant that the distance of the star from Earth could

be estimated. With this insight, Henrietta Leavitt gave us a ruler with which to measure distances in deep space.

We have to imagine astronomy in that era to appreciate the transformative power of Leavitt's discovery. From the time of Galileo to Pickering, people observed the sky and saw the planets, nebulae, and fuzzy patches of light with ever-increasing clarity. But the central questions remained. How big is the universe? Is our own galaxy, the Milky Way, all there is?

No sooner had Leavitt proposed her idea in 1912 than other astronomers began to calibrate and apply it to the heavens. One Dutch scientist used Leavitt's ruler to measure the distances between individual stars. It gave him a big number. The galaxy is vast almost beyond imagination. Then Edwin Hubble, armed with Leavitt's idea, used the biggest telescope of the time to change our view of the universe almost overnight.

In 1918, Hubble, a Rhodes scholar and law student turned astronomer, deployed his enormous new Mount Wilson telescope to find one of the stars made famous by Leavitt. This star was special. It wasn't alone in the sky; it sat inside a cloud of gas, known as the Andromeda Nebula. When Hubble applied Leavitt's ruler to the star, he encountered a stunning fact: the star, in fact the whole nebula that contained it, was farther away from us than anything yet measured. The game changer came from the realization that this object was much more distant than any star in our own galaxy. This nebula was no cloud of gas; it was an entirely separate galaxy light-years from our own. With that observation, the Andromeda Nebula became the Andromeda Galaxy, and the world above our heads became vast and ancient almost beyond description.

Hubble, using the largest telescope of the day, mapped everything he could see with Leavitt's variable stars inside. The Andromeda and Milky Way Galaxies were only the tip of the iceberg. The heavens were filled with other galaxies composed of billions of stars. Many of the fuzzy patches of gas seen by observers for a century or more were really star clusters that lie far beyond our own galaxy. In a scientific age when people were grappling with the age of Earth, then thought to be on the order of 10 million to 100 million years old, the age and size of the universe revealed our planet to be just a minuscule speck in a vast universe composed of innumerable galaxies. These insights emerged as people learned to look at the sky in a new way.

Hubble applied another technique to measure objects in the sky. This one relied on an essential property of light. Light radiating from a source that is traveling toward us looks more blue than light traveling away, which looks more red. This color shift happens because light shares some features with waves. Individual waves emanating from a source moving closer to you will look more compressed than ones moving away. In the world of color, more closely spaced waves are on the blue end of the spectrum, more separated ones on the red. If Leavitt's technique was a ruler to measure distance in deep space, then the search for color shifts in light was a radar gun to measure speed.

With this tool, Hubble found a regularity: stars emit redshifted light. This could mean only one thing. The objects in the heavens are moving away from us, and the universe itself is expanding. This expansion is not a pell-mell scatter; the heavens are scattering from a common center. Wind things back in time, and all the matter in the sky was at some distant time occupying a central point.

Not everybody liked this new idea; in fact, some experts hated it. Rival theories for the origin of the universe abounded. A proponent for one of them poked fun at Hubble's by giving it the moniker "big bang." Lacking in Hubble's theory, or in any other for that matter, was direct evidence in the form of a smoking gun.

The major breakthrough was an incidental by-product of people's need to communicate with one another. With technological innovations in wireless technology and expanding international commerce and collaboration in the late 1950s came a demand for transmission of radio, TV, and other signals across the oceans. NASA devised a special satellite, code-named Echo 1, for this purpose. Looking like a large shiny metal balloon, it was meant to bounce signals transmitted from one part of Earth to another. The problem with this system was that the signals returning to Earth were often far too weak to interpret.

Working for AT&T's Bell Laboratories, at the time a utopia for scientists doing creative science, Arno Penzias and Robert Wilson were designing a radar dish to detect the extremely weak microwave signals reflected from NASA's Echo 1 satellite. They spent a considerable amount of time, money, and expertise to develop a specialized radar dish for the task. Then, in 1962, NASA launched Telstar, a satellite that doesn't passively bounce signals but relays them with a boost of its own. The bad news for Penzias and Wilson was that their dish was now useless for NASA.

The good news was that, now free of anyone else's priorities, Penzias and Wilson were able to turn the dish to their real goal—observing the radio waves that hit Earth from space. But their wonderful contraption was not up to the new job. The sensitivity, so essential for their gig with NASA, made the dish a nightmare to work with. It picked up all kinds of faint signals and noise, almost like persistent static on a TV.

Their efforts to remove the noise read today like an attempt to find and remove a fine needle from a shag rug. First they tried to filter out the signals produced by radios. No luck; interference remained. Then they cooled the detector to -270 degrees centigrade, a temperature at which molecules come close to stopping their movement. Still interference. They climbed inside the detector and found that birds had sullied the interior via their digestive processes. Wiping away the evidence of those encounters helped a bit, but the interference remained. This background noise was constant through day and night and was about one hundred times more than they would have expected.

Unknown to Penzias and Wilson, a set of Princeton scientists used computer models to make a conjecture. If there was a big bang, some of the energy should be remaining in the heavens, drifting like smoke from an explosion. With 13.7 billion years of cooling and expansion since the event, this radiation should be found everywhere and be of a particular wavelength. This was quite a specific quantitative prediction, and it offered no room for waffling. A friend showed Penzias and Wilson these papers, and immediately they saw the real meaning of their static interference. The background interference was not noise; it was a signal. And it was of the exact type predicted by theory. Penzias and Wilson had discovered the remnants of the big bang, a discovery that won them the Nobel Prize in 1978.

Being a fossil hunter, I dig in the ground to uncover relics. But every astronomer is a paleontologist of sorts. As Carl Sagan famously said, the light of the stars we see was formed in chemical reactions from a long time ago. The vastness of space means that starlight hitting our eyes is no artifact; it is the real deal— a visitor from a time before the birth of our species, even in some cases our planet itself. With such time travelers coming down to us each night, the trick to reconstructing our past comes from learning to see the light and radiation of stars in new ways.

For thousands of years, mankind considered itself the pinnacle of life's creation on a planet sitting in the center of the universe. Science changed that perception. Leavitt, Hubble, and others helped us see that we live near the margin of a vast galaxy, in a universe of galaxies, with our planet one of many worlds. Darwin and the biologists had their say too. Our entire species is but one little twig on an enormous tree of life filled with all life on Earth. But each discovery that moves us from the center of creation to some obscure corner brings an entirely new relation between us, other species, and the entire universe.

All the galaxies in the cosmos, like every creature on the planet, and every atom, molecule, and body on Earth are deeply connected. That connection begins at a single point 13.7 billion years ago.

STARS ARE BORN

As a species whose history has been in oceans, streams, and savanna plains, we humans have had our senses tuned to the chemical and physical world of land and water—to predators, prey, and mates we can see or hear. Nowhere in our history has there been a premium on the ability to perceive extra dimensions, times on the order of billions of years, or distances in a virtual infinity of light-years. To achieve these insights, we repurpose tools that served us so well in our terrestrial existence to new ends. Logic, creativity, and invention project our senses and ideas to the far reaches of time and space.

The physics of the point that existed 13.7 billion years ago is mostly beyond our imaginations, not to mention our conceptual tools. Gravity, electromagnetism—all the forces at work around us did not have an independent existence. Matter as we know it didn't exist either. With everything that would become the universe packed so tightly in one spot, there was an enormous amount of energy. In such a universe, the physics of small particles, quantum mechanics, and that of large bodies, general relativity, were somehow part of a single, overarching, and still unknown theory. Just what that theory is awaits the next Einstein.

By about .0000000000000000000000000000000000000000001 second the universe was roughly 1,000,000,000,000,000,000,000, 000,000,000,000,000,000,000,000,000 degrees Fahrenheit, and the state of things starts to come more clearly into focus. This time begins the period of very rapid expansion of the universe. The big bang is not like an explosion where objects are projected from each other; space itself expands. With this expansion comes cooling over time. As the universe cooled and expanded, the forces and particles that make our world today emerged.

Einstein's relation $E = mc^2$ holds a key to the early events of the universe. The equation reveals the relationship between energy (E) and mass (m). Since the speed of light (c) is a huge number, it takes an enormous amount of energy to make an ounce of mass. The converse is also true: an infinitesimal amount of mass can be converted into a vast amount of energy.

One-trillionth of a second after the big bang, the universe was the size of a baseball. The energy contained in the universe at these early moments was the raw material for the production of a gargantuan amount of mass. As space expanded, energy, following Einstein's equation, converted into mass, in this case ephemeral particles. In such a hot and small universe, everything was unstable: particles formed, collided, and disintegrated only to repeat the process trillions upon trillions of times.

The particles at this moment of history were of two opposing kinds, matter and antimatter. Matter and antimatter are opposites and annihilate each other on contact. As energy converted to mass, no sooner were matter and antimatter particles produced than they collided. Most of these collisions led to the particles being completely extinguished. If this were the complete state of affairs, we—people, Earth, even the Milky Way—would never be. Particles would have been destroyed almost as soon as they formed. A slight—and by that we mean about one-billionth of 1 percent—excess of matter over antimatter was enough for matter to take hold in the universe. Because of that tiny imbalance, we are, as the physicist Lawrence Krauss once described, every bit the direct descendants of that one-billionth of 1 percent surplus of matter over antimatter as we are of our own grandparents.

At one second, our universe started to form entities we would recognize, if only very briefly. These are the collection of sub-atomic particles that make momentary appearances in some of the largest atom smashers today—leptons, bosons, quarks, and their kin.

A little over three minutes after the birth of the universe began the stirrings of one of the deepest patterns in the world, captured by the chart that is the source of either awe or angst for young science students—the periodic table. The periodic table catalogs all known elements by the weight of their nuclei. The chart drawn for this moment of time would be a huge relief to our students. There would be only three boxes on it: hydrogen, helium, and lithium.

Hydrogen and helium today remain the most common elements in the universe. Hydrogen makes up about 90 percent of all matter, helium about 5 percent. All of the others that compose us and run through the lives of people and stars are but a rounding error.

After 300,000 years the universe had cooled and expanded enough so that true atoms could exist. Nuclei were able to pull electrons into their orbits. This new combination of electrons with atomic nuclei set the stage for reactions that underpin every moment of our lives today.

We live in a daily marketplace of electrons, with trades measured in millionths of seconds. I write this book and you read it based on the energy released from these exchanges. The molecules in our bodies exchange these tiny charged particles as part of the daily business of their interactions. Some electron movements release energy; reactions involving oxygen tend toward this outcome. Other reactions serve to bind atoms into molecules or molecules with one another. These daily trades define the reactions between the planet's atmosphere, its climate, and the metabolisms of every creature on Earth. When you eat an apple, electrons from that material course through your cells to drive the metabolism to power your body. The electrons inside the apple to begin with were derived from the minerals in the ground and the water that fell from the sky. The electrons in both have cycled through our world for eons. And all of these came about well before the formation of the planet, the solar system, or even the stars.

With expansion and cooling, the stage was set: particles came together to make nuclei, nuclei came together with electrons to make atoms, and different atoms could now make the trades that are so essential for assembling ever-larger entities. One important thing had yet to take hold: gravity.

About 1 million years after the big bang, the universe cooled and expanded to the point where matter could get big enough for the force of gravity to have a meaningful impact on the shape of things. Order and pattern in the heavens emerge via a balance of forces: gravity serves to attract objects, while other forces, such as heat, and more mysterious ones, such as dark energy, serve to repel them. These relationships define the origin of the patterns we see in the universe, from the shape of gas clouds and stars to galaxies and planets. More fundamentally, they explain how chemistry itself evolved from a periodic table with only three elements to the one with over one hundred we live with today.

How did the world of atoms that make our planet and our bodies come about from the three that existed 13.69 billion years ago?

The march up the periodic table, from lighter elements like hydrogen and helium to heavier ones like oxygen and carbon, happens by the manufacture of ever-bigger nuclei. Under the right conditions two small nuclei can come together and make a larger one. The arithmetic of this combination depends on the physics of the nuclei themselves. In most cases, 1 + 1 does not equal 2: nuclei do not come together to make a new nucleus that is their simple sum. Often the new nucleus is lighter than that sum, and matter has been lost. But we know from Einstein's $E = mc^2$ that matter is not really lost; it is converted to energy.

These fusion reactions, then, can release enormous amounts of energy.

Humankind has tried to marshal the energies of fusion, but under normal circumstances atomic nuclei don't fuse spontaneously. The reaction takes a lot of energy to jump-start. Using this principle, Edward Teller, the father of the hydrogen bomb, made the first fusion device by attaching an atom bomb to another machine that allowed for the combination of nuclei. Atom bombs release energy by fission, a reaction that doesn't require much energy at the start. Teller, with his colleague Stanislaw Ulam, designed a system, code-named Ivy Mike, that was about the size of a small factory on the Pacific island of Enewetak. When it exploded in November 1952, the energy from the atom bomb forced the hydrogen atoms in the reactor to fuse, and a massive explosion ensued. Teller witnessed it from the seismograph in the basement of the geology building at the University of California at Berkeley. Enewetak was totally denuded, with a hole a mile wide in its center. Fragments from the island's lush coral reefs were ejected fifteen miles away. In analyzing the detritus left from the conflagration, the scientific teams discovered that the energy caused a number of large nuclei in the neighborhood to fuse, thereby producing entirely new elements never before seen on the planet. They were given the names einsteinium and fermium, after the scientists whose breakthroughs told us of the energy inside the atom.

Fusion reactions are the atomic engine that fuels the heat of stars. There is an essential difference between the Teller-Ulam device and celestial objects: Teller used an atom bomb to jump-start his fusion reactions, while the reactions inside stars depend on the force of gravity.

We can see evidence of these kinds of reactions today. Stare long enough at the constellation Orion using your peripheral vision, concentrating on the three stars that make the dagger on its belt, and if weather permits, you will see the fuzzy patch known as the Orion Nebula. When seen through a telescope, the nebula gains texture and complexity, appearing as a broad cloud with a number of smaller stars inside. The nebula itself is a huge field of gas, which, not entirely unlike that of the primordial universe, is giving birth to stars—about seven hundred of them. Of course, given the distance of the nebula from us, we are looking at baby pictures of starry infants from thousands of years ago.

During the formation of stars, fields of gas get so massive that the more particles they pull in, the stronger the force of gravitational attraction grows inside the cloud. At some point the mass of the gas cloud crosses a critical transition, and the gravitational attraction becomes a runaway process in which all the gas begins to collapse into a central point. Gravity pulls all the nuclei of the elements together, merging them. This union forces the nucleus to make a new combination; instead of one proton, it now forms a heavier nucleus with two. But this new nucleus is lighter than the sum of its parts. The lost mass, following $E = mc^2$, is converted to an enormous amount of energy released into space.

The size and life of any given star are defined by the push and pull that goes on inside the star: the force of gravity pulls elements in, and the heat of the fusion reactions works to separate things.

Stars are like an engine that first consumes one fuel, then, as this fuel is depleted, begins consuming a new one. The most basic star is one that fuses the smallest atom, hydrogen, to make helium. The sun is one of these ordinary stars. Over time, as hydrogen is consumed and the conditions become right, the star shifts to fusing the helium it made. For a while, it chugs along consuming the nuclei of helium to make even heavier elements. Once the helium is depleted, fusion reactions consume those heavier elements. And so on. This process leads to the production of oxygen, carbon, and heavier atoms. Through the fusion reactions inside stars, the periodic table went from having only three elements to having scores of them.

Fusion reactions in stars make most of the heavier elements in our bodies.

Stars can consume ever-heavier atomic fuels until they hit a stopping point defined by the laws of physics and chemistry. That point—the element iron—holds a very special place in the periodic table. Elements smaller than iron can fuse and concomitantly release enormous amounts of energy. Elements larger than iron can also fuse but, because of the structure of their atomic nuclei, not as much energy is released. More energy needs to be put into fusing these larger nuclei than can be gained from the fusion reaction itself. If, for example, iron formed the basis for a power company's nuclear reactor, less energy would be gained from the reactor than was put into it.

This equation is losing math for a star, but a huge gain for

We recycle. Hydrogen inside us comes from the big bang. Other elements come from stars and supernovae. And there they will return when the elements that compose us get spread around the universe by a future supernova.

us. As a star consumes all of the lighter elements, and marches ever higher in the periodic table in the fuels it consumes, iron accumulates in the center. As more and more iron accumulates, the fuel for fusion is consumed, nuclear fusion reactions cease, and the star begins to emit less heat. Iron nuclei, under the right conditions, can absorb energy, almost like a nuclear explosion in reverse. With so much energy released only to be absorbed, these conditions can set off a massive chain reaction that ends as a vast and catastrophic explosion. In seconds, these explosions release more energy than stars like our sun emit in their entire lifetime.

This blast is one kind of supernova (another kind can be triggered by collisions of stars). Supernovae work something like Teller and Ulam's crude device. The energy of one explosion brings new kinds of fusion reactions. Recall those fusion reactions for elements heavier than iron? Supernovae release so much energy that these expensive reactions happen. All the elements heavier than iron, such as the cobalt and cesium in our bodies, derive from supernovae.

Here comes the important part, at least for us. The blast of the supernova spreads atoms of the dead star across the galaxies. Supernovae are one engine that powers the movement of atoms from one star system to another.

The smallest parts of our bodies have a history as big as the universe itself. Beginning as energy that converted to matter, the hydrogen atoms originated soon after the big bang and later recombined to form ever-larger atoms in stars and supernovae.

The sky, like a thriving forest, continually recycles matter. With the heavens so full of stars manufacturing elements, then occasionally exploding and releasing them, only to recombine them again as a new star forms, the atoms that reach our planet have been the denizens of innumerable other suns. Each galaxy, star, or person is the temporary owner of particles that have passed through the births and deaths of entities across vast reaches of time and space. The particles that make us have traveled billions of years across the universe; long after we and our planet are gone, they will be a part of other worlds.

CHAPTER THREE

LUCKY STARS

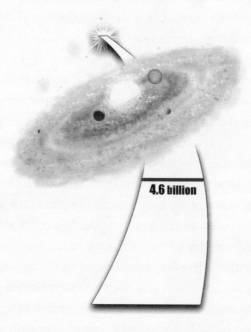

4.6 billion

E ver since the big bang, innumerable stars and galaxies have emerged and disappeared. We are relative newcomers to this party. By "we" I mean our entire solar system.

It took big ideas and big science to see how our little patch of the universe came into being. The Swedish thinker Emanuel Swedenborg was occupied by important questions throughout his life. Born in 1688, he lived most of his eight decades believing he should have one great idea per day. In his early years, he worked as a natural philosopher seeking to intuit the structure of the natural world. He inferred, for example, the presence of nerves and a nervous system. Turning his thoughts to the cosmos, Swedenborg proposed a theory for the origin of the solar system. He envisioned that the sun developed from a cloud of gas and dust that collapsed on itself and condensed. As the sun took shape, the primordial dust remained as a disk of debris that swirled around the young star. Over time, portions of this cloud coalesced to form the planets of the solar system. The idea was to remain dormant until two decades later, in 1755, when the philosopher Immanuel Kant had his go at developing ideas on the origin of the solar system. The theory he ultimately developed was largely similar to Swedenborg's.

Pierre-Simon Laplace (1749–1827) was one of the greatest mathematicians of all time, called by some the French Newton. Laplace's name peppers the fields of mathematics and statistics. There are, for example, the Laplace equation, the Laplacian operator, and the Laplace transform, tools to understand electricity, magnetism, and the motion of bodies in space. His real passion was to uncover the order in the heavens, the shape of the planets and the orbits of celestial bodies. With this intellectual goal, he converted the philosophical ideas of Swedenborg and Kant to the precise language of mathematics.

If a dust cloud in space gets to the right size, Laplace conjectured, particles inside will interact such that gravity will pull them together as other forces act to separate them. This push-pull means that a relatively amorphous cloud of dust can, under conditions where the pull wins out, develop into a swirling disk of debris. Over time, the gravitational attraction of the particles of dust in the disk break it into separate concentric rings—imagine a striped Frisbee. If the mass of dust in the rings is large enough, the particles could then condense to form the various planets of the solar system. These big events would happen not overnight but over timescales of millions of years.

Laplace's mathematical reformulation of Swedenborg's and Kant's ideas served as midwife for their transformation from interesting concepts to testable predictions. But the problem was that the technology to make the necessary measurements did not exist in the late eighteenth and early nineteenth centuries. Consequently, our understanding of the formation of the solar system stagnated for over a hundred years.

Enter big science. In 1983, scientists from the Netherlands, Britain, and the United States developed a satellite that could map the stars from an orbit around Earth. This predecessor to the Hubble Space Telescope was designed to perform one kind of observation really well: measure the infrared spectrum of the entire sky to assess how much heat is emanating from different stars. Through the course of their lives, stars emit everything from visible light to infrared, ultraviolet, and gamma rays. Our eyes sense only a small fraction of the light stars create, so astronomers use a wide range of telescopes, each tuned to different wavelengths of light, to capture a more complete view.

Because infrared signals from deep space are often weak, every source of interference needs to be removed from the sensors, even those made by vibrating atoms. To still the atoms, the device was cooled by liquid helium to a temperature of -452 degrees Fahrenheit. With room on board for only one year's supply of the coolant, the whole project became a race against time. It did its job, and the satellite, now defunct, continues to orbit the sky. In the years since, a small community of scientists has proposed a mission to give the satellite a helium recharge to put the sensors back in business. Limited budgets and the development of better technology have kept the satellite switched off.

Despite the short life span of the satellite's detectors, the mission was a huge success. In less than a year it charted almost 96 percent of the sky. The satellite mapped new asteroids and comets until, in early 1984, it captured a glimpse of a star radiating far too much heat for its size and type. We have a good idea about how much heat different kinds of stars should produce, and something was clearly different about this star. The source of that extra radiation became clear upon closer inspection of the images. The star was encircled by a vast cloud of dust and debris

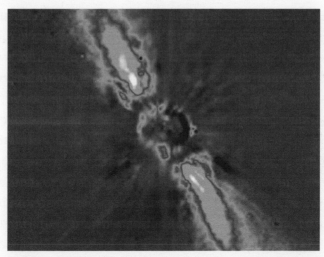

Beta Pictoris. One of the first images of a distant solar system being born.

that held heat. This system, Beta Pictoris, became the first example of a solar system caught in the act of being born. A prediction born as intuition and converted to mathematics was confirmed after two hundred years.

Soon after its formation, our solar system would have looked like Beta Pictoris. This moment of our history was chaotic; rocky debris fragments of different sizes collided with one another as they swirled around the sun. The gravitational pull of the sun meant that heavier material would orbit closer to it, while lighter particles and gas orbited farther away. To some extent, this state of affairs remains in effect today, with the solar system composed of rocky inner planets, Mercury, Venus, Earth, and Mars, and gaseous outer ones, Jupiter, Saturn, Uranus, and Neptune.

Whether the object of a search is Easter eggs, fossil bones, or a new kind of solar system, one discovery typically leads to the next. What once was rare turns up everywhere, often right under our noses. The years since the recognition of the dust surrounding Beta Pictoris have witnessed the launch of new satellites, the construction of ever-bigger telescopes, and the use of powerful computers to crunch all the data returning to Earth. This technology has changed our view of the heavens. Far from being a lonely solar system, ours is only one of many in the galaxy. The sky is filled with other worlds at different stages of their development surrounded by planets of almost every description.

Powerful technology and great ideas have transformed our notions of the heavens. But do not discount the impact of pure luck.

In the wee hours of the morning on February 8, 1969, a massive fireball woke residents of the Mexican state of Chihuahua. A visitor from space had arrived: a large meteorite that broke apart in the atmosphere. After learning of the event, scientists and collectors poured into the area in droves. Given the size of the boom, the collectors had expected a bonanza, but they had no idea of the extent until they looked carefully inside the rock. Tiny white patches interrupted the dull gray body of the rock itself. Meteorites with these specks were known before, but they were incredibly rare. Laboratory work on the few other meteorites with inclusions like these revealed grains that hint at the chemical signature of primordial rocks of the solar system.

The meteorite exploded into fragments that spread over about twenty-five square miles of desert. Two to three tons of fragments have been collected in the years since the impact. Even today, more than forty years later, pieces are occasionally found.

The impact could not have occurred at a more opportune time. In 1969, Project Apollo was in high gear. With *Apollo 8* having circled the moon just two months before the meteor strike and another as-yet-undetermined Apollo mission set to land on it, labs across the country were gearing up to investigate the chemistry of moon rocks. Now, at no expense to the taxpayer, special rocks from space had arrived right on our doorstep. Not only that, but the meteor was so huge prior to breaking up that there were a large number of fragments to share among the different chemistry laboratories capable of making sense of them.

Scientists performed the routine analysis of the atoms inside the rocks. Some of the mineral grains are so similar to those of Earth rocks that they point to a shared history of the bodies of our solar system, just as Swedenborg, Kant, and Laplace predicted. Other minerals can be dated using the decay of the atoms inside as a kind of clock. When a mineral forms, the atoms come together as a crystal structure. Once born as a crystal, some of the atoms, such as uranium and lead, change at a regular pace as defined by the laws of physics and chemistry. If you know the relative abundances of the different forms of the atom inside the mineral, and the rates at which they convert to one another, then you can calculate the time since the mineral formed (see Further Reading and Notes for more details). Uranium 238 converts into lead 206 very slowly; it takes 4.47 billion years for half the original amount to decay in this way. This slow rate of atomic change makes uranium and lead ideal atoms to measure the age of very ancient crystals. The uranium and lead concentrations of the Mexican meteorite point to an age for when the solar system got its start: 4.67 billion years ago.

But what was happening on Earth during these early moments? Direct evidence is hard to come by. In the ideal world, we would have a rock that formed at the moment Earth's crust cooled and has lain undisturbed for the billions of years since. The easiest geological conditions to study are those in which one layer of rock lies on top of the next, much like a birthday cake. The deepest layers would be the object of the hunt because typically those would be the most ancient. You could drill a deep core for them, but this is way too expensive for the typical geology budget. What's more, the drilling would be a bit of a shot in the dark: it would be hard to know where to look miles under the surface of Earth. You'd be better served to find places where the ancient rock layers are poking out at the surface of the crust. The challenge with finding these places is that the surface of the planet is continually being reworked. Mountains and oceans rise and fall. Under the action of such a dynamic planet, rock layers are buried, heated, and then eroded by water and wind. If the ideal geological conditions are regular cakelike layers, imagine a cake pulled asunder, crushed, and then superheated. Now throw 99.99999 percent of that dessert away. The hunger you'd experience trying to eat that cake would be similar to that of geologists who seek to find artifacts of the planet's formation.

Some places just feel primordial, almost like an ancient landscape frozen in time. At the Jack Hills in the arid desert of Western Australia, lowland scrub pokes out from orange and yellow bluffs of rock. Aboriginal art lies etched on boulders, the artists having died tens of thousands of years ago. The region's climate is so hot and dry that the nearby bays and inlets of Shark Bay are home to odd doorknob-shaped mats of microbes. These microbial communities are some of the most ancient living relics on Earth, with their closest relatives fossils that are over 2 billion years old. Fittingly, the bluffs of rock that jut to the surface match like jigsaw puzzle pieces to ancient ones buried deeply elsewhere. These are old-looking rocks too; heavily transformed by heat and pressure over time, they carry their history like wrinkles on a face. These rocks from the geological basement have been witness to most of the entire history of our planet.

Befitting such survivors, these layers have experienced eons of torment; from formation inside hot volcanic fluids to great pressures as they lay buried underground, finally to the stresses and strains that came when the layers were wrenched to the surface. Moment after moment is recorded in these layers; the trick, as always, comes from learning how to see the history inside.

Every rock in the ground is an artifact that, when you know how to interpret it, becomes a time capsule, a thermostat, even a barometer of the health of our planet. To wrest these details from stones, we have to zoom from a bird's-eye view of rock layers all the way down to a microscopic one. The smallest components of rocks—the individual grains of sand or minerals inside—often tell the biggest stories. One of these grains, zircon, has unique properties. It is virtually indestructible, and it can survive super-heating, high pressure, erosion, and virtually every other torture that the planet can throw at it.

Large, clear crystals of zircon make great fake diamonds. To those interested in the formation of the planet, zircons are far more valuable than gems, because zircon's durability makes it an ideal window into the ancient Earth. The rocks that contain zircons can come and go, but zircons are (nearly) forever. The clocks of uranium and lead from the Jack Hills produce a range of ages from 4.0 to 4.4 billion years.

The chemistry of zircons tells us more than the age of Earth. It holds a true surprise. The abundances of the various forms of oxygen inside the crystal can only have come from rock that interacted with liquid water as it formed.

Tiny grains tell the story of the solar system: it got its start over 4.6 billion years ago, and by at least 4.1 billion years ago liquid water, so essential for life, was already on Earth.

MAKING A SPLASH

We may live on the "blue planet"—unique in the known universe for its abundance of liquid water—but our bodies' ocean lies on the inside. Adult humans are about 57 percent water by weight. Our body dries out with every passing year; newborns are about 75 percent water, not much different from an average potato. Most of the body's water is not in the fluid of our blood, but remains locked inside the cells of our muscles, brains, and hearts. Metabolism of food and oxygen depends on water, as do the growth and communication of our cells. Even reproduction, with the motility of sperm and egg, is based in a fluid medium. Virtually every chemical reaction in our bodies depends in some way on the presence of water.

We are tied to water for more than our present lives: our bodies contain the history of water itself. The first 2.7 billion years
of our history was entirely in water, and the imprint is in every
organ system of our bodies. The fundamental organization of our
head is based on a series of swellings that develop into the bones
of our jaws, ears, and throats as well as the muscles, nerves, and
arteries that supply all of them. Equivalent structures are seen
in everything with a head, including fish and sharks. In these
creatures, the bones develop into the structures that support and
supply the gills. In a sense, the muscles, nerves, and bones that
we use to talk, chew, and hear correspond to the gill bones of our
fishy ancestors. This deep tie to gills is also seen in fossils, where
we can follow the transformation of gill bones to structures deep
within our own heads, including our ear bones.

While most of our past lay inside the water, the most recent
300 million years has been defined by our separation from it.
Our kidneys have developed specializations to help balance the
water and salts inside the body in the face of life on dry land.
Our reproduction doesn't depend as much on water as it did for
our ancestors: sperm and egg are fertilized inside the body, and
the developing fetus is shielded from the outside world by membranes and vessels that protect it and attach it to the mother. Our
hands and legs, structures so adept at supporting life on land, are
modified fish fins. Our terrestrial existence comes about through
repurposed organs that fish use to live in water.

The human kidney, like that of other mammals, is a magnificent adaptation to life on land; kidneys help kangaroo rats
and antelope live in dry deserts, surviving only on the water
locked inside the molecules of their food. Yet deep within this
most unique of terrestrial organs lie roots of its aquatic origins.
All jawless fish—ones we share a common ancestor with over
500 million years ago—have a very primitive kind of kidney:
tissues that run the length of the body, take fluid wastes from
the bloodstream, and dump them directly into the body cavity,

ultimately allowing for excretion from an opening at the tail. Bony fish, which share a common ancestor with us 450 million years ago, have a more specialized arrangement in which these clumps of tissue connect to a plumbing system that carries wastes outside the body. The most recent of these kidneys, the system that mammals use, doesn't run the length of the body but sits at the level of the lower back.

During our time in the womb, we form three different kinds of kidneys, one after the other. The first kidneys are clumps of tissue that line the body and open to the body cavity, much like those seen in jawless fish. The second, like those of bony fish, run the length of the back to a common plumbing system. The adult kidney, which appears at the end of the first trimester, replaces both of these. In our first three months, we track our fishy past.

Life's connection with water is no accident; the water molecule itself has special properties. With one oxygen and two hydrogen atoms, it looks something like a Mickey Mouse head: small hydrogen atoms form the ears atop a head made by a large oxygen atom. This whole molecule is polarized, with a negative charge at one end, where the oxygen resides, and positive charges at the opposite end, corresponding to the hydrogens. This arrangement makes water the ideal medium in which to dissolve a large variety of substances. Salts, proteins, amino acids—so many compounds can be incorporated into water that it provides the matrix for the chemical reactions on which life depends. No longer dependent on the vagaries of the water outside our bodies for our metabolic processes, we maintain that stable watery environment inside us.

Water has another property, one seen in a kitchen: it can exist as a liquid, solid, and gas within a relatively narrow range of temperatures and pressures. We have so many different kinds of interactions with water because it occurs on the planet as solid

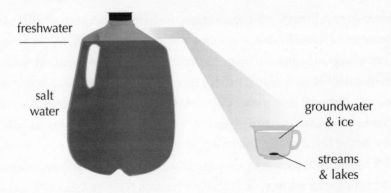

freshwater

salt
water

groundwater
& ice

streams
& lakes

Where's the water? The relative abundance of water on the planet.

ice, gas in the air, and the fluids that are the substrate for living processes. Over 97 percent of the planet's water lies in the
oceans, with the rest in the clouds, ice, and freshwater, and each
of these forms is vital to our existence and that of the planet.

Just as water is the matrix for the chemical processes that run
inside our bodies, so too is it for the metabolism of the planet.
Water raining from the sky and from melting ice erodes rock on
land and, as it flows from high to low elevations, returns minerals
to the sea. This gradual weathering provides the counterpoint to
the uplift of mountains and plateaus over geological time. Molecules in the air, many of which impact climate and atmosphere,
are continually recycled between rock and sea by the action of
water. Water provides the links that define a livable Earth.

The water inside our bodies and in the oceans also tells of its
origins. Being two parts hydrogen and one part oxygen, water
can be thought of as a two-to-one ratio of atomic nuclei derived
from the big bang to those derived from fusion reactions inside
stars. While their constituent atoms have a history that extends
across the universe, water molecules themselves are linked to the
solar system. The chemistry of the water in Earth's oceans, particularly the mix between different kinds of hydrogen atoms, is
distinctive and can be compared to the ice in comets, asteroids,
and other planets. Probes sampling water in the ice of the comet

Hale-Bopp, which passed by Earth in 1997, revealed differences between Hale-Bopp's water and that of Earth. This discovery was a huge disappointment to many because the reigning dogma in the 1990s was that comets were the likely source of Earth's water. Fans of the cometary hypothesis were in for a treat in 2011, when newer probes sent to other comets, such as Hartley 2, revealed water with very oceanlike proportions of atoms. The story of water is more complex than simply comets, because the more we look across the solar system, the more water we find. With powerful telescopes and ever-newer satellites, we have seen water turn up on the moon and within asteroids. There are even hints of water in the most unlikely places imaginable. Mercury is the closest planet to the sun; its surface reaches temperatures of 800 degrees Fahrenheit, hot enough to melt lead. NASA's MESSENGER satellite, sent to Mercury in 2004, captured photos of structures that have the distinctive reflective properties of ice deep within craters at the poles of the planet. Water may survive there because the craters of Mercury, shielded from the sun and on a planet with no atmosphere, are likely very cold. With so much water across the solar system, it seems likely that some water arrived here from space, but it is also possible that some came from the rocks of the forming Earth itself. When rock is superheated, as was the likely condition 4.5 billion years ago, it can vaporize and release water molecules trapped inside its molecular structure. Whether originating from the ice of comets or vaporized from the rocky debris of the early solar system, or both, each glass of water we drink is derived from sources at least as old as the solar system itself. And, as zircons tell us, water has been here on Earth in liquid form since at least 4 billion years ago.

Our history has been shaped by water, our existence made possible by it, and our future likely defined by our relationship to it.

Events far and wide have conspired to define our watery existence and with it, the fundamental structure of our bodies.

EVIL TWINS

Exhausted following a three-day meeting in California, I slumped on a sofa in a hotel lobby waiting for an airport shuttle bus. Seated across from me was an eminent colleague, his face partially hidden by the computer he had open on his lap. His facial expressions drew my attention. He was staring at his laptop, alternately laughing to himself and shaking his head in disbelief. I felt guilty watching him, so I tried staring at my bags to avert my gaze. Noticing my failed attempts to be discreet, he kindly beckoned me over to look at his computer. On the screen was a cliff face with a surface I had seen many times before. The way the layers crisscrossed is characteristic of rocks that formed in ancient dunes. I was familiar with this pattern, having seen it during fossil-hunting trips to Canada and Africa. I had even found fossils in this kind of rock. The rock beckoned; paleontologists dream of these kinds of geological exposures. But these photos were not from Earth. My colleague was part of the scientific team analyzing images returned from one of the Mars rovers, *Spirit,* and the computer images had just beamed back to Earth the day before.

In the 1988 movie *Twins,* the character played by Arnold Schwarzenegger is a virtual superman who goes looking for his long-lost brother. He ultimately finds his twin in a character played by Danny DeVito, a short, ungifted brother with a criminal past. They were born of the same mother, but an accident of fate left one brother with the gifts, the other with considerably less. Arnold's character, after meeting his brother, learns much about himself. So too can we learn when we look to neighbors in the solar system—Venus, Mars, and Jupiter—for insights into our planet and even the makings of our bodies. We have Arnold in our past and Danny in our future.

Separated at birth? Sandstones from the American West (left) and Mars (right).

For thousands of years, humans have looked to the sky for answers about life, time, and our place in the universe. Telescopes have enhanced our view, revealing moons on distant planets and canals on Mars. For the past forty years, we've sent hundreds of craft to the moon, asteroids, other planets and their moons, even deep space beyond the gravitational pull of our sun. *Apollo 8* propelled the first humans beyond the gravitational pull of Earth to enter one dominated by another celestial body. Circling the moon on Christmas Eve 1968, William Anders captured the rise of Earth over the surface of the moon. About twenty-five years later, the unmanned *Voyager* spacecraft began to leave our solar system, departing from the gravitational pull of our sun to enter deep space. Engineers turned the cameras back to reveal Earth. What was for *Voyager* a single pixel in space, and for *Apollo 8* a globe, was a blue oasis of water and air in a world unique among all known ones in the universe.

Even before Project Apollo, observations of Venus changed the way we see our place in the universe. The bright planet looks like a sphere, but next time you have the opportunity, scan it with binoculars or a telescope. What you will see is something that nearly got Galileo executed when he first interpreted it in 1610. Venus, like our moon, has phases that extend from crescent to full and back again. From these kinds of observations, Galileo was able to prove that the planets, including our own, rotate around the sun rather than Earth.

Being near the size of Earth, and relatively close to the sun, Venus has long been thought to be our closest planetary relative—so much so that the earliest interplanetary missions were sent there in the hopes of finding life. Some scientists even thought that when we landed on Venus, we would discover a tropical world, almost like that of Earth during the age of the dinosaurs.

The first hint that something strange was happening on Venus came in the 1930s, when observers looked at the planet with a new kind of telescope. Rather than measure the intensity of light, this telescope, at the Mount Wilson Observatory in California, deconstructed the light into its electromagnetic spectrum. The spectral pattern hinted that Venus's atmosphere was composed of 99 percent carbon dioxide.

In 1962, Venus won the lottery as the first extraterrestrial planet to receive visitors from Earth when NASA planners launched the first spacecraft of the Mariner project. This mission was a huge undertaking; liftoffs are always dangerous, but in 1962 they were particularly so. *Mariner 1*'s liftoff went so awry that the destruct button had to be pushed to prevent a disaster for the towns of coastal Florida. The mission that followed, *Mariner 2*, only had the capability of carrying about forty pounds for all the scientific measurements it would perform. After a successful liftoff, *Mariner 2*'s trip to Venus took about three and a half months. The small number of instruments that the probe carried were to make some very large discoveries. They revealed that Venus has surface temperatures that were roiling hot, about

900 degrees Fahrenheit. Venus has surface pressures about ninety times those of Earth; you would need to go half a mile underwater to feel that kind of pressure. And the instruments confirmed that the atmosphere is almost entirely carbon dioxide. *Mariner 2* found that our close planetary relative, our near twin in size, is most similar to hell.

How could such a dead ringer for Earth in so many ways be so different? Part of the answer would come from new kinds of probes.

In the 1960s, while NASA was gunning for the moon, the Russians were developing machines to land on Venus. Getting a mission to actually visit and record data on Venus is a tricky business. These machines need to be lightweight so that they can be put aloft, but therein lies a huge trade-off: the immense pressures of the planet's atmosphere give precious little time to make measurements before the probe is crushed like a beer can at a football game. Not surprisingly, the list of early missions reads like a list of calamities. *Venera 1* lost contact en route. *Venera 2* lost contact upon arrival. *Venera 3* crash-landed on the planet. *Venera 4* entered the atmosphere, sent a few signals, then was lost. But persistence paid off. *Venera 9*, launched fourteen years after *Venera 1*, landed on Venus and sent back the first grainy black-and-white photographs. Subsequent missions landed and were able to analyze soil samples and the environment. What did they find? Venus has thunder and lightning. Venus has lava rocks much like those of Earth. Venus may be a hot, high-pressure, and carbon-dioxide-rich world, but it is strangely similar to our own planet.

Then came NASA's *Pioneer* mission, launched in 1978. This probe was a miniature science lab in space, carrying equipment that could measure the composition of the clouds and the chemical constituents of the atmosphere, among other things. When *Pioneer* entered a Venusian cloud, some of the sulfuric acid inside touched one of the devices. The device could then look at the atoms inside, particularly the different kinds of hydrogen. The ratio of different atoms of hydrogen in a sample of gas is influenced by the presence of liquid water. In the measurement of these atoms came a surprise. Venus is dry as a bone today, but at some point in the very distant past it had oceans.

Venus and Earth were born twins, but our fates have been completely different thus far—Venus lost its water, while our planet kept it. Venus's relatively close position to the sun defines a world where liquid water cannot be maintained. The loss of water may be behind many of the differences between the two planets. On Earth, water facilitates the removal of carbon dioxide from the atmosphere through a long chain of chemical interactions with rocks. These reactions are not possible on Venus. Having no liquid water, Venus is like a closed container being pumped with gas; with volcanoes spewing carbon, and no way of removing it, pressure just builds over time. As a result, the planet gets hotter and hotter. Venus is in a runaway greenhouse, set up by its loss of water.

Our neighbor on the opposite side of our planet's orbit from the sun, Mars, tells a different story. On Mars, we have yet to find active volcanoes spewing gases, lava floes, or a moving crust. Canyons and canals carry the signature of having been sculpted by flowing water. Dormant volcanoes dot the surface. If there was liquid water, then there needed to be temperatures in the ranges we experience here on Earth. The surfaces that reveal extensive flowing water are scarred by time, often pockmarked by small and large impacts that may have happened billions of years ago. Recent probes reveal seasonal flowing water on Mars today, but these flows are a far cry from those that created the deep canyons that exist on the planet. Mars's vigorous and aquatic past is largely frozen in time.

Many of the differences between Venus and Mars derive from the heat balance of the planets. Venus lost its water because, being close to the sun, its water evaporated and set off a runaway chain reaction of ever-increasing heat. Since Mars is relatively far from the sun, it likely did not receive enough heat to sustain liquid water. Mars's relatively small size also contributed to its loss of heat. All else being equal, small entities have more surface area for their size than do big ones. For example, children have relatively more surface than adults. More surface area means more loss of heat, so children start to shiver in a cold pool quicker than adults do. Planets are no different. Mars's shivers led it to lose its heat and much of its geological activity.

In the planet business, being in the right place, at the right size, with the right materials, at the right time is everything. We live on a planet that is habitable because it formed at just the right distance from the sun, in a gravitational balance with its neighbors, with just the right amount of material to support a world with liquid water, recycling crust, and an atmosphere. What can we thank for being the lucky recipients of such an inheritance?

BY JOVE

For the Romans, Jupiter was the god who oversaw oaths and laws and thereby defined the social balance that maintains a just society. The planet Jupiter plays an analogous role in the physical and biological worlds.

Jupiter has two and a half times the mass of the rest of the planets in our solar system combined. Its mass is over three hundred times that of Earth; more than eleven hundred Earths could fit inside Jupiter. This colossus exerts, through its gravitational pull, an enormous effect on its neighbors. It sucks asteroids and comets into its field. As for the planets, Jupiter competes with the sun to pull them into its orbital plane. This cosmic tug-of-war defines our own orbit and has guided much of our history.

Over 4.6 billion years ago, as dust swirled the star that would become our sun, clumps of debris formed, as Swedenborg,

Kant, and Laplace envisioned. Jupiter was the equivalent of the two-ton gorilla in the crowd: being the most massive planet in the solar system, it had a profound impact on its neighbors. The gravitational attraction that produced the planets made them, in effect, compete and interact with one another. Imagine the tugs felt on the forming Earth everywhere from the sun, from other planets, and from the center of attraction inside the young planet itself. A huge planet, such as Jupiter, with a proportionally large gravitational field around it, influenced how much material was available for Earth to form and where it would lie relative to the sun.

Computer simulations of the origin of the solar system suggest that Jupiter formed before Earth. Competition with Jupiter for debris meant that the position of Jupiter defined the shape of the rest of the solar system. If Jupiter formed closer to the sun, it would have led to fewer but larger rocky planets in the interior of the solar system. If Jupiter formed farther from the sun, there would likely have been a larger number of smaller planets. Our planet's mass and its distance from the sun—the benevolent conditions that have supported liquid water and life—are due in no small part to the influence of Jupiter.

Our dependence on Jupiter lies in every part of our being, from the presence of liquid water on the planet to the size, shape, and workings of our bodies. The formation of Jupiter defined the size of Earth and, in so doing, the pull of gravity on all things on its surface. A simple thought experiment reveals the web of interconnections. If Jupiter had formed closer to the sun, then Earth would have been larger and heavier and the pull of gravity experienced by Earth's denizens greater. Even in the unlikely event that such a strange Earth managed to hold liquid water, life on the planet would be very different. As every engineer knows, if you want to make a beam resistant to bending with the same material, just make it relatively wider. All else being equal, a heavier Earth means fat and wide bodies to cope with the greater tug of gravity. Conversely, a smaller Earth would have meant less tug of gravity on evolving bodies; hence, their proportions would need to be longer and lighter. The mass of Earth defines the gravity we experience and, in so doing, controls virtually every aspect of our lives, from our size and shape to how we move about, feed, and interact with the planet.

From the imbalance of matter over antimatter in the moments after the big bang, and the formation of Jupiter that defined our livable planet, to the way a single sperm out of millions fertilized the egg that defined our genes, it is easy to celebrate the lottery each of us has won to be here on a habitable planet. But it is a virtual certainty that within the next billion years the sun will run through its hydrogen fuel, expand, and become superhot. In the process, Earth will almost certainly lose its water. The subsequent loss of water will cause a runaway greenhouse effect, superheating the surface of our planet. Earth will become like

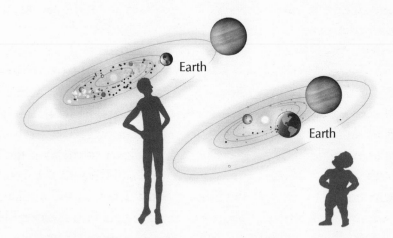

Alternate fates? Seeing the effect of Jupiter on the shape of our bodies is almost like looking into fun-house mirrors. We would have had more elongated bodies and lived on a smaller planet if Jupiter formed farther from the sun (left), and been short and squat if it formed closer in (right).

Venus. The next planet with liquid water, and the conditions for life, will likely lie farther from the sun. Perhaps it will be a more distant body in our solar system that currently has ice—one of the moons of Jupiter such as Europa, or Enceladus, a moon of Saturn. Our good fortune, the perfection of circumstances that have defined our existence, is just a moment in time.

CHAPTER FOUR

ABOUT TIME

4.5 billion

A trip in a time machine to Earth of 4.5 billion years ago would not only be eerie; it would be perilous. With an atmosphere lacking free oxygen and raining acid, you'd need a space suit far beyond the technology of modern science to survive. Impact after impact of rock and ice from space made the surface sometimes roil at thousands of degrees Fahrenheit. With this heat, there were no oceans: liquid water may have formed a few different times, only to evaporate away. As a break from this desolation, you might hope for beautiful moonlit nights. Forget about it. There was no moon.

Artifacts of the transformation of this primordial world into our modern one are strewn across different bodies of the solar system. Six missions landed on the moon and returned samples to Earth. Carrying mini geological kits, astronauts collected rocks from craters, highlands, and lowlands of the lunar surface. The specimens are today stored in liquid nitrogen in repositories in Houston and San Antonio. A number of small moon fragments have been given as gifts to foreign dignitaries, while others grace public exhibits. The bulk of the rocks, about 850 pounds in all, remain to be studied. The few samples that have made it to labs tell important stories of the origin of our world.

One of the biggest lessons from moon rocks is how normal many of them are. In terms of mineral content and structure, moon rocks are more similar to those on Earth than others in the solar system. One similarity is particularly telling. Oxygen atoms can exist in different forms, defined by the number of neutrons in the nucleus. By measuring the neutron-heavy and neutron-light versions of oxygen in any rock, a very informative ratio can be calculated. Each body in the solar system carries a unique chemical signature written in the proportion of different versions of oxygen in their rocks. The reason is that the oxygen content inside a planet's rocks is sensitive to its distance from the sun when it formed. The oxygen composition of moon rocks, though, is virtually identical to those of Earth. This means that the moon and Earth formed at the same distance from the sun— perhaps in the same orbit.

With all of these similarities, there remains one very significant difference between moon rocks and those of Earth. Moon rocks almost entirely lack one class of elements, the so-called volatiles. These elements—nitrogen, sulfur, and hydrogen—share one important geological fact: they tend to vaporize when things get hot (hence the name volatiles). Some great event in the distant past must have baked the moon rocks, releasing their volatiles.

The lessons of the moon rocks are clear—the minerals on the moon formed at the same orbital distance from the sun as Earth and then suffered some kind of blast. What do these facts tell us of the origin of the moon?

The current theory for the formation of the moon envisions something like a cosmic demolition derby. In these automotive mosh pits, common at fairgrounds in the 1970s, cars intentionally smashed into one another, with the last car running being the winner. Along the way, cars would slam into each other with wild abandon. The most violent of these collisions would eject the light outer layers of the cars, hubcaps and bumpers, leaving the inner ones hopelessly entangled.

This type of collision offers an insight into how the Earth-moon system came about. Over 4.5 billion years ago, a large, perhaps Mars-sized, asteroid is thought to have collided with the forming Earth. Much like the twisted mélange of car parts in a demolition derby crash, the collision ejected lighter parts of each body while the heavier pieces fused. The lighter debris, consisting of dust and smaller particles, now depleted of volatile elements, began to orbit Earth as a disk. Over time, this debris disk coalesced as the moon. The cores of the two bodies did not propel into space but liquefied under the great heat of the impact, only later to cool and solidify as the new core of Earth. In addition, the impact so whacked Earth that it left a 23.5-degree tilt in its axis of rotation.

Initially, there were two large bodies in the same orbit of the sun. Then they collided, forming what we know as Earth and moon today. Ever since that impact, the two bodies have been locked in an orbital dance—Earth and moon exert gravitational pull on each other, while the laws of physics and momentum tie the speed of the spinning of Earth to the rotation of the moon. The impact on our lives is as straightforward as it is profound: the length of days and of months, like the workings of the seasons, derive from the Earth-moon system. Every clock and

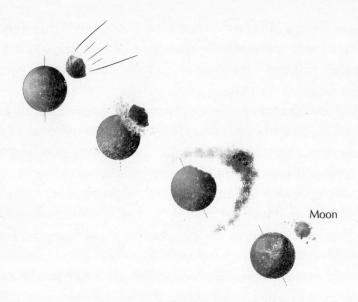

The big whack. The origin of the moon.

calendar, like the cells of our bodies, holds artifacts of a cataclysm that took place over 4.5 billion years ago.

KEEPING TIME

The Romans had an effective way of controlling troublesome officials in the far-flung regions of their empire. Instead of gerrymandering districts to stay in power—to help friends and get rid of foes—Caesar and his cronies found the ultimate way to retain control. They gerrymandered the calendar. Have a political friend in one region? Add a few extra days to his term. Want to get rid of a foe in another place? Lop days of his rule off the year. This was wonderfully effective; however, over time, not only did the decentralized calendar make ruling difficult, but the year became a patchwork of political kludges, fixes, and compromises.

The nature of Earth's rotations in space makes it ripe for these kinds of abuses. We all learn this material in school, but most of us forget the meaning of the planet's rotations by the time we are in college. A recent survey of Harvard undergraduates asked the simple question: What causes the seasons? Over 90 percent of them got the answer totally wrong. The answer has nothing to do with the amount of light that hits Earth during winter and summer, nor with Earth rocking back and forth, nor with the planet getting closer to the sun over the course of the year.

As we've known since the days of Copernicus and his contemporaries, the moon rotates around Earth, while Earth retains its constant 23.5-degree tilt as it rotates around the sun. The angle that sunlight hits the planet changes at different parts of the orbit. Direct light generates the long days and heat of summer; tilted and less direct light gives us shorter and colder winter days. The seasons aren't generated by Earth rocking back and forth; they derive from the planet having a constant tilt as it rotates around the sun.

Because of the different orbits that affect our lives—ours around the sun and the moon around us—there are choices to make when constructing a calendar. Of course, the length of a year is based on the rotation of Earth around the sun. If we know the longest and shortest days, we can carve up the year into months based on the seasons. Another way to do this is to base the calendar on the position of the moon as it goes from full to partial to new every twenty-nine days. The problem is that you can't synchronize a lunar calendar with a seasonal, or solar, one. The number of lunar cycles does not correspond easily to the number of seasonal ones.

So what do we do? We add fudge factors. Julius Caesar's calendar had a leap year every three years to keep the months in line with the seasons. The problem with this calendar for the Catholic Church was the extent to which the date of Easter wandered. To rectify this situation, Pope Gregory XIII initiated a new calendar in 1582. Italy, Spain, and a few other countries launched it immediately following the papal bull, resetting October 4, 1582, to October 15, 1582, losing eleven days. Other countries followed to different degrees. Britain and the colonies, for example, only accepted it in 1752. One of the most important issues to iron out, naturally, was when to collect taxes.

Years, months, and days can, at least in theory, be based on celestial realities, but minutes and seconds are mostly conventions. Our calendar has seven days because of the biblical story of a six-day creation, followed by a day of rest. Minutes and seconds are in units of 60 due to a matter of convenience. The ancient Babylonians had a number system based on 60. It turns out that 60 is a wonderful number because it is divisible by 1, 2, 3, 4, 5, and 6.

Humans are a timekeeping species, and much of our history can be traced to the ways we parse the moments of our lives. These intervals are based as much on astronomical cycles as on our needs, desires, and the ways we interact with one another. When the necessities of shelter, hunting, and survival were highly dependent on days and seasons, humans used timepieces derived from the sun, moon, and stars. Other early timepieces relied on gravity, with hourglasses that used sand or water clocks such as those first seen in Egypt in 4000 B.C. Our need to keep time has itself evolved; an ever-increasing necessity to fragment time corresponds to the demands of our society, commerce, and travel. The concept of moments parsed into seconds would have been as alien to our cave-dwelling ancestors as seeing a jet plane.

There are clocks in our world that do not rely on convention, political choice, or economic necessity. The DNA in our bodies can serve as a kind of timepiece. Averaged over long periods of time, changes to some parts of the DNA sequence happen at a relatively constant rate. This means that if you compare the DNA structure of two species, you can estimate how long ago they shared an ancestor, because the more different the strands of DNA from two species are, the more time they have changed separately. As we've seen with zircons, atoms in rocks also tell time. Knowing the ratio of different versions of the elements uranium, argon, and lead can tell us how long ago the minerals in the rock crystallized.

The different clocks in bodies and in rocks don't tick independently; they are part of the same planetary and solar metronome. Comparisons of the DNA inside humans, animals, and bacteria speak of a common ancestor of all three that lived over 3 billion years ago. This is roughly the age of the earliest fossil-containing rocks. The broad match of dates from rocks and DNA is all the more remarkable given how the rocks have been heated and heaved over the same billions of years that DNA has mutated, evolved, and been swapped among species. Agreement between these different kinds of natural clocks leads to confidence in our hypotheses. On the other hand, discordance between the clocks in DNA and those in rocks can also be the source of new predictions. Whale origins are a case in point. With some of the largest species on the planet, blowholes in the middle of their heads, ears specialized for a form of sonar, and odd limbs, backs, and tails, whales are among the most extreme animals on Earth. Yet, as observers have known for centuries, their closest relatives are mammals: they have hair, mammary glands, and innumerable other mammalian affinities. But which mammals are their closest relatives, and when did whales enter the seas? Comparison of the DNA of whales with that of other mammals revealed that whales likely diverged from odd-toed ungulates such as hippos and deer. The differences in the genes and proteins implied that the split happened nearly 55 million years ago. But this created a whole new puzzle for paleontologists. Not only were there no fossils that showed transitional organs in the shift; there was nothing that ancient with whalelike features in the fossil record. The gap served as a challenge. Vigorous paleontological exploration brought confirmation: the discovery of whale skeletons with ankle bones similar to those of hippos and their relatives inside rocks over 50 million years old. And it all happened by relating the different clocks in rocks and DNA.

Rocks and bodies contain more than clocks: they also hold calendars. Slice a coral, and you will find that the walls of the skeleton are layered with light and dark bands. As they grow, corals add layers of mineral to their skeletons, almost like slapping plaster on a wall. The ways that the mineral forms depend on sunlight, so the variation in the layers reflects the waxing and waning of each passing day. Mineral growth is fastest in summer, when the days are long, and slowest in winter, when the days are short. Consequently, bands deposited in summer months will be thicker than those at other times of the year. Count the number of layers embedded in each cycle of thick and thin layers, and what do you find? Three hundred sixty-five of them. Coral skeletons can be an almanac of days of the year.

The beauty of corals lies not only in the reefs that reveal the splendor of the underwater world but in the insights they give us into our past. Crack rocks along the sides of roads in Iowa, Texas, even north into Canada, and you will see coral reefs that once thrived in ancient seas hundreds of millions of years ago. The city of Chicago is built upon an ancient coral reef. And reefs like these tell the story of how time itself has changed. Go to fossil reefs 400 million years old, and you will find four hundred layers inside the corals—suggesting that each year was actually four hundred days long and contained a whopping thirty-five more days than our current year. What accounts for this discrepancy? Since the duration of a year is fixed by Earth's rotation about the sun, the days must have been shorter 400 million years ago than they are today. To make the algebra work, each day had to have been twenty-two hours in length. In the eons since those corals were formed, two hours have been added to every day.

Like a slowing top, Earth spins slower and slower with each passing moment, making days longer now than in the past. As the planet rotates, the water in the oceans moves about and serves to brake the spin of the planet. That is why today is two milliseconds longer than yesterday.

Fossil corals are silent witnesses to the lengthening of days. Clocks and calendars abound in the natural world, sometimes in the most surprising places.

IT IS IN YOUR HEAD

In the rush of pitching my tent, I inadvertently left a hummock of tundra under the floor. With a mound in the center and slick nylon surfaces inside, my sleeping bag slid to one corner each time I drifted to sleep. After a frustrating few hours of writhing like a pupa in a cocoon, I became determined to find a flat surface, and in a fit of fatigue and desperation I jury-rigged one by contorting myself over heaps of clothes, books, and field gear. It was a good thing we expended a lot of energy that first day setting up camp; my exhaustion led me to a reasonable facsimile of sleep.

I arose to the bright morning sun and dressed quickly, not wanting to hold the team back. Today would be our first day in Greenland looking for fossils, and the excitement made me surprisingly alert despite the fitful rest.

I made my way to the kitchen tent, my first goal being to get the coffee going. Our field gear was packed so tightly for the trip north that simply finding the breakfast containers was no small task. After about ten minutes of fumbling with the packing lists and crates, I broke out some cooking supplies and got the java brewing.

Life was good. It was a clear, bright Arctic summer morning. The dry air made images incredibly sharp; features in the distance looked as if they were right next door even though they were miles away. Warming my fingers against the coffee mug, and relishing the stillness, I walked in my mind through the different hills I was going to hike that day.

After a few cups and about twenty minutes of savoring the calm, I realized something was wrong. The world was still, a bit too much so. With each passing minute of silence, I began to feel more alone.

A glance at the clock revealed the cause for my solitude: it was 2:00 a.m. Yet here I sat, fully dressed, primed for a whole new day, and bristling with energy. I felt like a total chump, albeit a well-caffeinated one. Returning to sleep was an impossibility, so I broke out a novel I was saving for a snowy day and struggled to read for the next few hours until my companions arose.

It was the light, of course. The walls of my tent did not block it out, leaving the inside illuminated at all hours. My brain, acclimated to the southern world, was completely in tune with the equation "light equals day and dark equals night." Because that simple relationship was lacking in the twenty-four-hour daylight of Arctic summer, my brain's usual cues were utterly useless. My sleeping colleagues, old hands at fieldwork, prepared by bringing eyeshades, while all I had was a flashlight.

Those first few days were a real jangle. I felt off-kilter, as if the insides of my body were struggling to keep up with a whole new planet. Think of a major case of jet lag, but without any night whatsoever, the only reference point comes from a clock. The longer I dwelled in the landscape, though, the more my brain became attuned to it. The sun traces a large ellipse through the sky, casting different shadows throughout the day. Almost without thinking, the brain begins to make a sundial out of any standing object. Of course, in the high Arctic we lack trees; any large rock or tent ends up doing the job.

From jet travel we all know that our sleeping and wakefulness are matched to the sun. Virtually every part of us—every organ, tissue, and cell inside—is set to a rhythm of day and night. Kidneys slow down at night. That's a wonderful trait if you want to minimize trips outside bed—something very useful when inside a sleeping bag in the Arctic. Body temperatures vary over the course of the day, with the coolest ones happening at 3:00 a.m. Liver function is time dependent as well: the human liver works slowest in the morning hours, meaning the cheapest dates would be at breakfast.

Our bodies respond to more than days; they also are tied to seasons. The changes from winter to summer bring new patterns of light, temperature, and rainfall. Animals are tied to these in the ways they feed and reproduce, and humans are no different. Even our moods relate to the season. By some estimates 1.4 percent of Floridians suffer from seasonal affective disorder, compared with 14 percent of New Hampshire residents.

Drunks see time flying by, with the party just getting going as everybody leaves. Cannabis brings an eternity to a twenty-minute episode of *The Three Stooges*. Intense concentration or emotions make us lose track of time. Even the proverbial "watched pot" that never boils is a statement about how our perception of time is sometimes at odds with the clock itself.

In 1963, a young French geologist had a plan to change the way we think about time. By the age of twenty-three, Michel Siffre had visited some of the largest unexplored patches of Earth. These were underground, and by mapping the world below, Siffre revealed vast caverns and glaciers inside the Alps. The subterranean landscape is a beautiful and dark world, and in this void Siffre was inspired to ask a whole new question.

What happens when people completely disconnect from the clock? Each of us is a slave to it: we chop our days into little moments and plan our lives around them. Not only do we live in a world defined by natural time—the dark of night and the light of day, the warmth of summer and the cold of winter—but we have inserted man-made inventions into this equation. Beepers, buzzers, and alarms tether us to each passing moment. What happens when we completely cut the cord that binds us to these stimuli?

Siffre intended to be his own lab rat and concocted a plan to live for two months in a cavern two hundred feet belowground, completely removed from normal human existence. He would bring food, a sleeping cot, and artificial light, but—and this is the important point—no timepiece or anything that could even indirectly give a clue as to time. Siffre's only connection to the outside world was a telephone with which he called his friends on the surface to inform them of the times he spent awake and asleep. The plan was for a sixty-day disconnect from the normal light-dark cycles of our world and from our clocks that are based on them.

A meticulous note taker, Siffre dutifully recorded each passing day on a calendar with his bodily functions and mental states. His diaries record his daily movements, his body temperature, his mood, and his libido.

On the thirty-seventh day of his records, with twenty-three days yet to go, Siffre was on the phone with a colleague from above. Pierre, one of his chums, asked, "How much time in advance do you want to be warned that your experiment is about to end?"

"At least two days to gather up my things."

"Start getting ready," Pierre responded. The experiment was over. By relying solely on his mental clock, Siffre had lost twenty-three days.

What happened?

One answer lay inside Siffre's diaries. Having recorded when he woke and went to sleep, he called his friends when he was able so that they could log the real time for him. But lacking a watch or alarm clock, he had no idea how long each interval of sleep really was. What he perceived as brief ten-minute catnaps were in reality eight-hour slumbers.

His misperception of time ran deep. At one point in the experiment, Siffre called his friends to see if he could mark off two minutes simply by counting. Most of us can pace this off roughly, within ten seconds or so. Siffre began counting from 1 to 120, in an attempt to march off the seconds in two minutes. This simple task took him five minutes.

When the team crunched the numbers in Siffre's diary, they came to a fundamental realization. The hours of Siffre's biological day, when he slept and was awake, were not some wandering, random affair. The number of hours he slumbered and those he was awake almost always totaled close to twenty-four. A rest-activity cycle of twenty-four hours is a close approximation of life on the surface. Siffre's perception of time underground was completely off, but his bodily cycles marched along at an earthly pace.

Word of Siffre's feat ignited a fad of isolation research. In the years since his time in the cave, volunteers have endured sensory deprivation while gizmos recorded their vital signs, brain activity, and behaviors. Some have sat for weeks or months in chambers with light either restricted or tightly controlled. Other people have attempted the truly extreme, like the sculptor who entered an isolation experiment in a chamber with the goal of living in complete darkness for months. That experiment had to be scuttled after only a few days when he started to lose his grip on reality.

Through all of this sensory deprivation, one consistent pattern emerges. Many of the biological urges we experience—sleep, hunger, and sexual cravings—cycle at a regular pace regardless of whether we live in a dark cave, room, or any other isolated environment. Time exists on the clock, in our perceptions of it, and somewhere deep inside us.

Curt Richter had an inauspicious beginning to a scientific career. After a stint in the army during World War I, he entered the Johns Hopkins University to explore the extent to which animal behaviors are based on inborn instinct. He arrived in Baltimore in 1919 to work with a senior scientist already famous for his work in the field. Unknown to Richter, his new adviser had an offbeat way of training students. Richter was given the usual trappings of a graduate student's existence: a little office in which to study, a library card, and some supplies. Once established with his kit, Richter was left completely on his own. Each day consisted of no set meetings, no required courses, no seminars; it was an existence with no structure whatsoever. There would be no babying, for this was a sink-or-swim introduction to a scientific career.

Soon after Richter established himself in Baltimore, his adviser handed him a cage with twelve normal rats inside. His instruction to Richter was as simple as it was intimidating: "Do a good piece of research."

Richter began by feeding the rats some bread and staring at them for days on end. Like a good scientist, he started to record details about their daily lives: when they ate and what they did. Then one day he made the observation that changed his career and ultimately launched an entire new field of science. As he reminisced years later, when describing his rats, "They just jumped around the cage for long periods and then were quiet again."

The rats clearly had set times of activity and rest and of hunger and satiety. Richter began to fiddle with things a bit. Once he had a reasonable idea of their activities, he started to leave the lights on in the lab overnight. Then he would turn the lights off for days on end. The activity patterns of Richter's rats stayed on track. The rats responded in the way the caveman Siffre did years later.

With that simple observation, Richter saw in these rats the

question that was to become his lifelong obsession: What was the basis for these innate daily rhythms? What controlled them?

Richter turned to studying blind rats. With no obvious cues to perceive light and dark, these animals retained set daily patterns of sleep and feeding. He postulated that there must be something inside, some clock that keeps things ticking along. If so, where in the body did it reside?

The first candidates were the glands that secrete hormones controlling heart rate, breathing, and other bodily functions. Richter gave the rats chemicals to disrupt their hormone balance, then altered their diets, and later he even took out the glands to see what happened. He performed more than two hundred of these kinds of experiments over the years. And in every case, he got the same result. No matter which part he removed, or which hormone was blocked, daily patterns of activity just ticked along normally.

Richter turned his attention to rats' brains and nerves. He systematically lopped off pieces of the rats' brains to see the effects. In 1967—forty-eight years after his adviser's gift of the cage—Richter removed one little patch of tissue at the base of the brain. Excision of a fleck of cells not much bigger than a grain of rice obliterated virtually all of the rhythmic behaviors in the rats. The rats' internal clocks were in a tiny field of cells right behind the eyes, later named Richter's patch.

How could a bunch of cells keep time? Answers to that question were to elude Richter and his brute-force approach of removing pieces of the body. The key came from an entirely different approach to science.

CLOCKS EVERYWHERE

Genetic mutants are the stuff of biology: six-toed cats, two-headed snakes, and conjoined goats are not just curiosities; they have the power to tell us the fundamentals of how bodies are built and how they work. While mutants tell us which genes have gone awry, their scientific importance lies in how they reveal what normal genes can do.

Let's say you find a mutant animal that lacks eyes. Clearly there is some genetic difference between mutant and normal animals that causes eyes to be missing. There is a lot of work to be done to isolate the actual genetic defect. But by doing breeding experiments between the mutants and the normal animals, you can ultimately identify the gene itself and, if you are really persistent, discover the actual stretch of DNA involved. Knowing the DNA is a key to understanding the molecular machinery that controls how eyes are built. The same is true with virtually every gene that acts to build bodies.

Scientists have scoured populations looking for mutants. Careers have been made, and Nobel Prizes won, by cataloging mutants or by finding just the right one with an extra toe, different jaw, or oddball eyes, limbs, or heart. While the rewards can be high, this approach is often like rolling the dice. Some mutations crop up only once in every 100,000 individuals. Unfortunately, the creatures we are most interested in, mammals like us, are the hardest to work with, because they take a longer time to develop than other creatures, they take more resources to rear, and they spend most of the critical phases of their development hidden from the outside world inside the female.

Because of these difficulties, for the past hundred years, flies have been the best animals in which to study genes. As opposed to mammals or reptiles, for example, you can get lots of flies: they reproduce and develop fast. With such a steady supply of embryos and an entire community of scientists doing the research and sharing data on them, an enormous number of different kinds of mutants have not only popped up in labs but been cataloged, described, and stored for others to study.

But why wait for nature to produce mutants when you can make them? The process of mutating DNA is relatively straight-forward, on paper at least. First, you take some flies and treat them with chemicals or radiation that interrupts the copying of their DNA. With altered DNA, a variety of mutants emerge in the next generation. In flies, the wait from one generation to the next is only about a day. Then scan the mutants for whatever trait you are interested in. Some will have bizarre legs, others mutant eyes, and so forth.

By the late 1960s, Seymour Benzer's laboratory at Caltech had become a hub for the study of mutant flies of all sorts. A young student, Ronald Konopka, entered with a plan and the desire to take some risks making mutants. Benzer's lab was inter-ested in looking at behaviors, and by the time Konopka set to work, Benzer's group had already accumulated mutants with odd reproductive and courtship dances.

Konopka had a plan to use this same approach to unravel the genetic workings of the clocks in the body. Smart money at the time was against his success. Most scientists thought the body's clock was just like a stopwatch and way too complex to dissect easily.

Konopka had come to the right place to make inroads into this problem. Seymour Benzer had a personal motive to understand body clocks. He was a notorious night owl, working late hours in the lab, yet his wife was in bed soon after dinner. Perhaps mutant flies could provide help to a marriage where the partners inter-acted with each other only at dinnertime.

Konopka spent what must have seemed like a near eternity looking at developing flies. Flies hatch from eggs as maggots, feed for a bit, then develop hard cases around their bodies from which they later emerge as metamorphosed flies. Flies emerge from their cases in early morning, at the time of day when it is normally coolest. This behavior is a manifestation of their inter-

Seymour Benzer.

nal clocks; flies reared in artificial light–dark cycles will always emerge at the end of a dark cycle, when their brain is wired to expect the cool of dawn.

After about two hundred tries at making mutants, Konopka found one batch of flies that were completely messed up: some emerged too early from their cases, others too late, and still others at random times during the day. The differences in time emergence almost certainly reflected some sort of genetic defect. Here was a defect that reflected a possible clock mutation.

The lab bred each type of fly selectively with another of the same type, making whole family lines of the different mutant flies. With these lines, Konopka and Benzer set the stage for understanding the molecular machinery that keeps time.

Like any good gene, Konopka's makes a protein that does the real work in the body. Knowing the gene means that you can ask the important question: Where and when is the protein active?

In normal flies the amount of protein peaks in the early night, then drops to almost nothing during the day. Knowing this normal rhythm allowed Konopka and Benzer to make sense of the mutants. In the early hatchers, the protein levels peaked early. In the late hatchers, they peaked late. And what about the proteins in the flies with no rhythm? No working protein. The activity of the protein matched exactly what you would expect from one that was behind the daily rhythm.

By now a number of other labs were digging into the problem. Using powerful technologies of fly genetics, they isolated the DNA and found a number of other genes that work in this system. With each new gene, a finer understanding of the fly's biological clock emerged.

A clock, even the biological one that Konopka and Benzer found, works something like a pendulum that swings back and forth at a set pace. The regular movement provides a basis to measure units of time. Looking at the clocks inside bodies means seeing molecular equivalents of pendulums inside cells. Here, the major timekeepers come about from chemical chain reactions in which each step occurs at a rate set by the laws of physics and chemistry. When DNA is activated, it makes interacting proteins that are transported through the cell and perform a number of tasks, one of which is to activate portions of DNA again, letting the cycle begin anew. This pendulum-like swing of chemical changes is defined by how fast proteins are made, combine, travel through the cell, and ultimately interact with DNA.

Mutant flies opened up not only a way to see genes that control biological clocks but also a new way to think about people.

About ten years after Konopka and Benzer's discovery, a graduate student, Martin Ralph, was setting off to see if there were genes in mammals that controlled body clocks. He and his adviser in Oregon were working with Syrian hamsters, running them on treadmills to monitor their activity. Normal hamsters had rest and activity times, measured by how long they ran on a treadmill, that totaled to about twenty-four hours.

When each new shipment of hamsters came to the lab, Ralph would run them on his treadmills, hoping for some oddball pattern of rest and activity to emerge. He was making a huge bet that a mutant would randomly show up in one of his weekly shipments.

One day Ralph took the newly received batch of hamsters, put them in the cages, and let them run. To his joy—and relief, no doubt—one individual had a total daily rest and activity cycle that was dramatically different from twenty-four hours. This one hamster functioned on a twenty-two-hour cycle. When he bred the hamster with others, he found the offspring also had the shorter daily rhythm of rest and activity. The aberrant clock of this hamster was caused by a mutant gene.

When Ralph and others looked in more detail at the activity of the gene, they saw that the major effect of the mutant was a little patch of cells inside the brain. The mutation affected cells inside Richter's patch. Ralph's bet had paid off.

The plot thickened in the early 1990s when a patient entered a sleep clinic in Salt Lake City complaining of an odd problem. She got a regular eight hours of sleep each night. But no matter how hard she tried to stay awake, she always fell asleep at 7:30 p.m. With a normal eight hours of sleep, that meant she was waking at 3:30 a.m. Her clock ticked at a normal pace, only everything was shifted. Sound familiar?

Entering a sleep clinic means being probed, outfitted with patches, and connected to machines as you try to slumber, all in an effort to put some numbers on bodily functions—breathing, heart rate, body temperature, and so on—during the night. Every test showed that there was nothing strange about this woman's health or behavior.

Shortly after the testing began, the patient revealed the critical fact that other family members shared the same sleeping timetable. There were several puzzles to the family's sleeping habits. Other members of her family rose very early, but they lived in different places. One fact after another led the researchers to suspect that the cause didn't lie in their homes, diets, or local environments.

The researchers assembled a full pedigree, mapping the family tree against which relatives rose early and which did not. Then the pattern emerged: here was a classic genetic trait, just like the fly mutants.

Looking at a family tree doesn't reveal the actual change to DNA itself. That level of understanding requires taking cells from the body, by swabbing the cheek of each family member, and comparing the DNA inside the cells between those who rise early and those who wake at more normal hours. If there is a genetic difference, it should be seen in some part of the DNA that consistently varies between the risers.

Isolating the DNA of the early risers and comparing it with that of the normal sleepers showed that the altered sleeping pattern was caused by a very precise change to one gene. Genes and proteins leave their fingerprints everywhere. Once you know the sequence and structure of a gene or protein, it is a relatively simple matter to search for them in other cells. This trail led from the DNA of mutant human early risers, to that of Martin Ralph's early running Syrian hamsters, ultimately to the gene that caused Konopka and Benzer's flies to hatch early. Each species has its own specialized protein interactions inside the clock, but the basic genes and principles that drive the molecular pendulum are all there. Flies, hamsters, and people share parts of their clocks, their genes, and their histories.

Similar clock DNA not only leads us to our deep connection to other species but also reveals a fundamental machine inside the cells of our body. The same kind of genetic clock is within each of them, from Richter's patch and the skin at the tip of your finger to those of the liver and brain. If you have a sleep disorder that has a genetic basis, the doctor can diagnose it from a scratch of skin, drop of blood, or swab of your cheek. Every cell ticks a daily rhythm with a molecular clock that has parts at least as ancient as animals themselves.

What sets the molecular clock inside our cells and aligns it to our days? A travel alarm clock might keep a twenty-four-hour rhythm, but it has no way of knowing what time zone it is in; it needs something to set it to where it is on the planet. Why do we experience jet lag or, in my case, rise at 2:00 a.m. in the Arctic summer?

Our cellular clock is tuned to the outside world by a number of triggers, the most important of which is light. Most of the light that enters our eyes ends up as a signal to the parts of our brain that interpret visual information. Some of these signals, though, get sent to a different part of the brain—to Richter's patch of cells. The path from Richter's patch travels to a little pea-sized gland at the base of your brain called the pineal. To some, including the great French philosopher René Descartes, the pineal is the seat of the soul. In some lizards and fish it actually forms a kind of third eye that records light information directly. In us, it is like a relay center for information. It emits the molecule melatonin, which triggers responses all over the body.

This reaction—from light to brain to the pineal gland to melatonin and its targets across the body—tunes our bodies to the light of the day and the darkness of night. When we travel to a different time zone, this pathway eventually resets us to a new regime of light and dark.

Shine bright light on somebody's eyes in the middle of the day and what happens? Usually nothing at all beyond the usual adjustments of his or her visual apparatus. Shine bright light in people's eyes at dusk, and you can affect sleep. People hit with

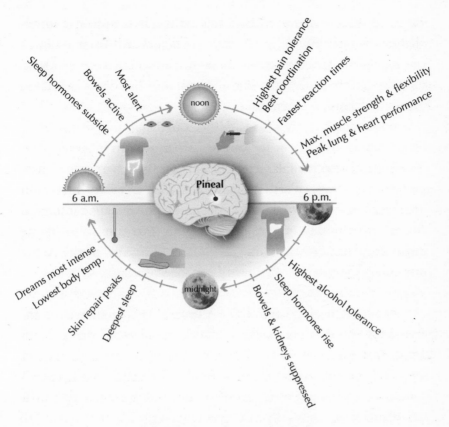

Body functions follow a diurnal clock.

light at dusk tend to become tired later than normal. The opposite is also often true: shine bright light on people at dawn, and their sleep cycle will be shifted earlier than normal. Our sleep cycle is dependent on light shined on us when our brains expect it to be dark. In a world of gadgets that blare artificial light into our eyes at all hours of the day, we are resetting our clocks with each text or e-mail sent in the middle of the night. We live in an age of disconnect between the ancient rhythms inside us and our modern life.

Much of our health depends on clocks: shift workers who sleep during the day and work at night have higher rates of heart disease and some cancers, notably of the breast. Researchers studying mice discovered that the error-correction machinery of the DNA of skin cells functions on a clock: it's most active in the evening. The DNA that gets copied in the morning is likely to carry the most errors. The UV radiation of the sun causes cancers in the skin when it induces errors in copying DNA. Putting these facts together leads to the conclusion that in mice UV light hitting the skin in the morning is more carcinogenic than evening and afternoon exposure. Humans also have these clocks, but ours are reversed relative to mice: our DNA error-correction apparatus is most active in the morning. This means tanning at the end of the day is more carcinogenic than doing so in the morning. Even our metabolism is affected by the clocks inside our cells: some kinds of obesity can be correlated to a lack of sleep.

Given that mechanisms of DNA function and cell division are dependent on internal clocks, it should come as no surprise that a number of medicines are most effective at certain times of the day, when our brain anticipates the level of light. Our susceptibility to disease, and our treatment of it, carries the deep signal of a planetary cataclysm that happened over 4.5 billion years ago.

Dotting the landscape of southern Indiana are cemeteries that house the graves of Europeans who settled in the region in the late eighteenth century. This was a hearty crowd whose harsh existence is recorded on their tombstones. Few lived past the age of forty, and judging by the carved dates, the cemeteries were busy places some years. By an accident of geography, these settlers found a near-perfect material for their headstones. The fine grains and hardness of each stone preserve etchings from the early nineteenth century as sharply today as when they were originally carved.

We are so accustomed to looking at the front of tombstones that it is easy to overlook other edges that have stories to tell.

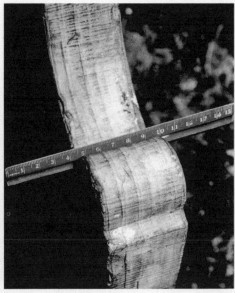

Tombstones from the Hindostan quarry (left) have ridges that correspond to the changing tides (right).

The sides of the pioneers' grave markers are not even; they are composed of a series of ridges with sharp edges and small depressions. The stones were quarried from a pit near Hindostan, Indiana, from an exposure that reveals how the rock was formed. Hundreds of millions of years ago, this part of Indiana lay under the sea. Year after year, sediment settled from the tidal waters, leaving small ripples in the mud. There is a rhythm to these marks, recording the variations of the tides throughout the year. As Earth spun and the moon circled it, the water rose and fell, only to be recorded as ridges in the sediment. The sides of the grave markers reveal the tidal rhythm from a moment when Earth rotated faster and the days were shorter than today. Time is sculpted everywhere on these tombstones, by the work of man and of the planet. The bodies in the graves and the rocks that mark them are united by the history they share with colliding and rotating celestial spheres.

THE ASCENT OF BIG

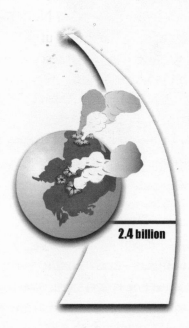

2.4 billion

As the young sun pulled matter into its orbit over 4.6 billion years ago, lumps of rock and ice smashed together and combined. The cataclysm that gave us the moon was only one of many. Judging from the ages of the craters on the moon and meteorites here on Earth, collisions were the order of the day until about 3.9 billion years ago. Then this violent start gave way to a long period of relative quiescence.

In this quiet Earth lay a scientific puzzle that confounded scientists. The very top, or youngest, layers held fossils—the different shells and bones that can be seen in any natural history museum today. Lying below were ancient layers of rock with no evidence of anything living inside: no fossil bones, markings of animals, or plant spores; no evidence of any living thing whatsoever.

This lifeless layer of basement rock wasn't just a sliver; the barren layers were miles deep. All of human history—and the entire known history of life on the planet itself—were confined to a thin veneer of crust. If the entire history of Earth were scaled to a year, with its formation on New Year's Day and the present being December 31, Earth was utterly lifeless until mid-November. Translate this relative timescale into years, and we find that about 4 billion years of the history of our planet was devoid of living things. To Charles Darwin, the abrupt appearance of life was an "inexplicable mystery."

The solution to Darwin's mystery, along with clues to how our modern world emerged, came from a source no one could have ever predicted.

The steelworks of Gary, Indiana, stand like hulking fossil skeletons of a thriving bygone age. In the 1950s, humming mills fed a burgeoning automotive industry, with plants sprinkled across the Midwest. The need for iron was great, and geologists such as Stanley Tyler worked to feed it by studying the iron-ore-containing rocks in the region.

Because the rocks that contain iron tend to be among the oldest rocks around, ore geologists like Tyler focus mostly on the geological basement. As Tyler knew, these rocks were great for industry, lousy for paleontology.

One afternoon in the mid-1950s, Tyler was studying samples he collected from a deep test pit in northern Michigan. Rock chips from different layers were brought back to his lab in Madison, Wisconsin, where each was ground thin and placed on a slide to observe the fine structure of minerals and grains. Sitting with a checklist and a microscope, Tyler performed the usual scoring and counting of the color, grain size, and mineral con-

Elso Barghoorn.

tent that are the necessary but rote measurements of much geological work.

When Tyler looked at one of the slides under the scope, he saw something familiar yet completely out of place: coal. He knew that coal reflects ancient plant detritus and that most of the coals known at that time were from layers no older than 350 million years, when plant life was abundant. But Tyler also knew the age of his sliver of rock from Michigan. It was almost 2 billion years old. Something was entirely wrong.

Not believing his own eyes, Tyler discreetly passed the rocks to experts. Shuffling from one specialist to the next, the samples eventually ended up in the hands of Elso Barghoorn, a curator of early plants at the Harvard Museum of Comparative Zoology. Scanning the slides under a microscope, Barghoorn immediately confirmed Tyler's hunch. Tyler had found the earliest life yet known on the planet—coal-forming algae and other microorganisms.

This answer led to a whole new puzzle. The more Barghoorn looked at Tyler's slides, the more ancient species he saw. Each sample was brimming with algae spores, filaments, and the remains of thousands of single-celled animals. The basement of rock on our planet wasn't devoid of life but teeming with it.

Now the "inexplicable mystery" lay not in the missing life but in understanding the plethora of living things that were once hidden from our view. When did life get its start on the planet? What did the earliest living things look like? Because there was a world of questions to ask and creatures to find, a new kind of paleontologist emerged—one whose life's goal was to recover microscopic fossils from rocks billions of years old.

Hunting for fossils in the first 2 billion years of Earth's history has special challenges. Rocks of the right type to hold super-ancient fossils are likely to have been eroded away, baked by internal heat, or transformed by various movements of Earth's crust. Let's say you actually manage to find ideal rocks. How can you tell that a microscopic bleb or filament was once a living thing and not some mineral or inclusion in the rock? The science of early life is one of constructing multiple lines of evidence, from the shape and structure of the putative fossils to the chemistry of the rocks that hold them. The hunt is for creatures that not only look like single-celled forms but also reveal the chemical workings of metabolisms.

With this playbook, Stanley Tyler, Elso Barghoorn, and their scientific descendants exposed the hidden reality inside the rocks: the earliest fossils are now known to be over 3.4 billion years old. Life arose early in the history of our planet and, once off to that start, expanded rapidly to become a menagerie of different kinds of bacteria, algae, and their relatives.

Despite their incredible diversity, the organisms that dominated the first billions of years of the history of our planet share one important thing. They are all single celled and microscopic. Some of them lump together to form colonies, but no individual dwelling on Earth for the first 3 billion years was larger than a grain of rice. Big, in the world of living things, had yet to come about.

THROUGH THE LOOKING GLASS

"That one's dead," Thomas Barbour, the director of Harvard's Museum of Comparative Zoology, shouted while considering a frog lying motionless on the grass in front of him.

On the roof above Barbour stood his colleague Professor Philip Darlington, holding a bucket of frogs with one hand. With the other, he was pitching the frogs one by one onto the lawn five stories below.

Barbour nodded as each frog hit and remained motionless in the grass below his feet. When Darlington descended with his empty bucket, he asked Barbour how the animals withstood the impact. Seeing the frogs strewn about, Barbour offered, "All dead."

Darlington was a naturalist of the old school: when he wasn't teaching courses at Harvard College, he was off in the jungle collecting new species, beetles in particular. Tales of his days in the field are legendary, including the time he was grabbed by a crocodile, pulled to the bottom of a stream, and, as the crocodile began to consume him, kicked himself free. Hiking miles to safety with shredded legs and hands, he wrote to his wife later that night only that he had an "episode with a crocodile."

In the midst of his explorations in the 1930s, debates were raging about how animals dispersed to new places around the globe. This was the era before plate tectonics, and there were two major ways to explain animal distributions: either there were land bridges between continents—now lost—that allowed animals to walk from place to place, or animals could be blown by wind, water, and storm. Darlington was a firm believer in the latter and his boss, Thomas Barbour, in the former.

The frog "experiment"—something we would obviously never perform today—began as an argument. During coffee at the museum one afternoon, the two got into a tussle about the theories and made a wager. Barbour held that dispersal by wind was impossible because animals would die upon impact. Darlington countered that wind dispersal would work over considerable distances for small animals. The two agreed on the rooftop test of the theory.

And what of Darlington's dead frogs?

Within minutes after the impact, each frog arose and hopped away. Soon the grass was filled with frogs bounding in different directions. Darlington proved his point.

Of course, there is nothing unique about frogs that allowed them to survive such a fall; this ability is a reflection of their size. Small animals accelerate more slowly during a fall than do large ones, since they experience more air resistance for a given mass. In describing this phenomenon, J. B. S. Haldane, one of the founders of evolutionary genetics, said, "You can drop a mouse down a thousand-yard mine shaft; and, on arriving at the bottom, it gets a slight shock and walks away. . . . A rat is killed, a man is broken, a horse splashes."

Let's say you want to predict what an animal can do—how long it lives, how it moves about—and what it looks like without ever having seen it. A number of factors may be influential: the kinds of foods the creature eats, where it lives, where it sits in the food chain, and so forth. People have explored this issue by cataloging measurable traits that creatures possess and hitting the data with a number of different statistical tools to gauge which measurement accounts for the differences we see. In analysis after analysis, one factor reigns supreme in its predictive power—size. Know a creature's size, and you can make educated guesses about much of its biology, including its resting heartbeats per minute (smaller animals have higher heart rates), its perception of danger (larger animals have less fear), even its life span (in general, larger means longer).

Virtually every part of the world we experience is influenced by our size, even how we visualize size itself. The size and shape of our pupils, eyeballs, and lenses influence our visual acuity just as the shape and structure of the different components of our ears affect the sound frequencies we hear. Because ours is a world tuned to the predators, prey, and other entities of our ancestors' worlds, we are like a radio that can receive only a small number of channels; vast portions of the world remain hidden to us. Extending our gaze beyond the limitations of our biology has meant seeing our size, and ourselves, in a brand-new light.

Anton van Leeuwenhoek (1632–1723) spent much of his career as a draper and found himself needing to develop magnifying glasses to assess the quality of his fabrics. Becoming fascinated by the properties of glass, he manufactured new kinds of lenses that magnified objects many times beyond the tools common to his trade at the time. He tweaked the shape of the glass again and again, each time seeing smaller things, ultimately magnifying objects two hundred times. With each new lens he crafted, he was exploring a new world.

Van Leeuwenhoek was famously secretive about how he crafted his lenses. For centuries it was thought that he polished the glass into ever-finer slivers. Then, in 1957, a science writer for *Scientific American* speculated on van Leeuwenhoek's trade secret: he made his lenses by heating glass rods and pulling them apart. Reheating the broken end made a little ball at the tip. When this little glass bead was separated from the rod, he mounted it in a mechanical contraption that held both the specimen and the bead at a set distance. Peering through the glass bead revealed its magnifying properties, and the bent glass served as a kind of lens.

Everything became fodder for van Leeuwenhoek's microscope. In one famous experiment he took the plaque from an older gentleman's mouth and put it in his scope. In it, van Leeuwenhoek found "an unbelievably great company of living animalcules, a-swimming more nimbly than any I had ever seen up to this time. . . . Moreover, the other animalcules were in such enormous numbers, that all the spittle . . . seemed to be alive." This is thought to be one of the earliest known descriptions of bacteria. He looked at pond water and found a carnival of life inside—from algae to microbes—and later described human semen as containing little tadpole-like creatures.

People flocked to see van Leeuwenhoek's cabinet of wonders in his house in Delft. There, they became the first humans to

Anton van Leeuwenhoek and his microscope.

catch a glimpse of a novel world. For thousands of years all of
human knowledge was centered on the universe we can hear,
touch, and see with our natural-born senses. By extending
beyond our biological inheritance, van Leeuwenhoek revealed
we are all big creatures living in a world chock-full of innumerable microscopic ones.

Just a few decades before van Leeuwenhoek's revelations with
a microscope, Galileo Galilei (1564–1642) was doing the exact
opposite: grinding glass to make a telescope. With the most powerful telescope of his day, equivalent to a large set of binoculars
from an outdoors store today, Galileo was able to see Venus's
phases, that Jupiter had moons rotating around it, and that huge
nebulae populate the sky.

Van Leeuwenhoek looked down through a microscope to find
a small world. Galileo looked up to the sky and revealed a huge
one, with incomprehensibly large planets and vast distances. In
van Leeuwenhoek's world, we are humbled by the diversity of
microscopic life beneath our noses and within our bodies; in
Galileo's, by the sheer size of the world around and above us.

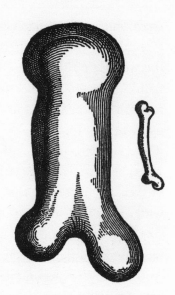

Galileo and his etching of the comparison of the leg bones of an elephant and a mouse.

How did this new humility come into being? By finding new ways to use glass.

Around 1633, over twenty years after Galileo constructed his telescope and described the rotation of the bodies of the solar system, he was found guilty of heresy and sentenced to be imprisoned for the remainder of his life. Because he was already seventy years old, he was placed under house arrest, first in Siena and later in his own home in Florence. During this period of confinement, Galileo spent about five years writing about physics. He was forbidden to publish in Italy, so a Dutch printer, Louis Elzevir, secreted his manuscript out of the country.

Galileo's book—different from any science exposition we are familiar with today—consists of a fictional conversation among three men who set out to explore the fundamental laws of the universe. Their conversation holds the beauty of mathematics applied to the world around us.

On day two of the confab, the three explore the laws that govern the shapes of all objects. What happens to objects when they get bigger? How do small objects differ from big ones? Think of trees: short trees can get by with relatively narrow trunks, but tall trees are an entirely different matter. Assuming the properties of the wood are the same, tall trees will need proportionally wider trunks to protect them from bending and breaking. This simple relation between size and shape defines much of the world around us. A lithograph of upper leg bones from Galileo's book reveals it all. A mouse femur and an elephant femur are similar in many ways, because they have the same joints, and they are composed of similar bones. But the elephant femur is proportionally much wider than that of the mouse. Just as with the tree trunks, larger size necessitates new shapes. This relationship holds for dinosaurs and elephants as much as it does for bridges and buildings. And the reason, as Galileo recognized, is because larger entities have to deal with ever-increasing effects of gravity.

Galileo envisioned that the gravitational pull defining the orbits of celestial bodies also has an effect on animal and plant organs. Bodies are pulled to Earth to a degree that is proportional to their mass. Heavier creatures, being pulled relatively more, need to change their shape to support themselves. This relationship even explains Darlington's rooftop experiment with frogs. Lighter animals accelerate less during a fall than do big ones for these same reasons. The force of gravity can mean the difference between life and death for large creatures like us.

Gravity is not a significant factor for creatures that dwell in van Leeuwenhoek's microscopic world. Look no further than at a fly or an ant on a wall. For the fly, the gravitational pull of Earth on its body is negligible; the forces that are really important are the ones that bind molecules together. A fly can stick to a wall because these sticky, and tiny, molecular forces are proportionally more significant than gravity for a light animal. Imagine a hippo on that wall: gravitational pull would far exceed the pull of the molecules on its feet. No amount of molecular Velcro would work to keep the hippo stuck to the side of a room.

We, relatively big creatures that we are, barely think of these molecular forces when we go through our day. We swim in water and feel that the water is more viscous than air. But if we were a small creature, say a bug no longer than a quarter of an inch, these forces would dominate our existence, and swimming in water would feel like swimming in Jell-O. The surface of the water would take on new meaning. At our weight, we can dive right through to the bottom of a pool. Since a number of these molecular forces are at work on the surface of fluids, a bug can walk on it.

In 1968, F. W. Went wrote a classic paper, published in *American Scientist*, that explored the consequences of size for humans. His starting point was the seemingly absurd question: Could an ant begin a workday like a person? Understand that Went was not a crank; the discoverer of a critical plant hormone, member of the august National Academy of Sciences, on the faculty of Caltech, and, later, director of the Missouri Botanical Garden, he was an eminent Renaissance man of science. In the obvious "no" to Went's ant question lie a number of profound biological truths. Went reasoned that an ant shower, of course, would be out of the question. The droplets of water, being bound by those molecular forces, would hit the ant's body like cannonballs. A morning cigarette (the essay was written before the Surgeon General's report had a major impact on our behavior) would also be impossible. The smallest effective size for a controlled fire is about that of the ant's body. Saying good-bye to the spouse and kids would be different too. The ability to hear deep tones—slow vibrations in air—is possible only at larger size. Also dependent on size is the ability to hold a job in the first place. The development of brain capacity for thinking, forethought, and memory requires a certain body girth. The ant story makes one point abundantly clear: many of our abilities, such as talking, using tools, designing machines, controlling fire, and so on, are possible only because of our size. Size defines the opportunities of our species.

Our ancestors gained new possibilities when they made the

shift from van Leeuwenhoek's microscopic world to the Galilean one over a billion years ago: they left a world dominated by intermolecular forces and entered one more influenced by gravity. This great moment of our past is written inside our cells, deep within the rocks, and in the ways many of us die.

IN THE AIR

Impressions of disks, ribbons, and fronds in slabs of 600-million-year-old rocks are a pretty unremarkable bunch of fossils. But looks are deceiving. These fossils reflect revolutionaries, a new kind of individual that the world had not yet seen. These are the first creatures with bodies composed of many cells.

The advent of bodies changed the planet forever. A single cell is restricted in size because molecules can only diffuse over short distances, a measurement set by the laws of physics. This limitation affects how an animal can feed, respire, even reproduce. Small animals can transport oxygen across their bodies by simple diffusion. Once animals get large, they need new mechanisms to move nutrients and wastes about. How do they deal with this? Large animals have specialized systems to circulate blood, carry wastes, and capture and pump oxygen to their far-flung cells. These kinds of specialized organs are game changers in the world of size. Hearts, gills, and lungs are all inventions necessary in large animals. This complex machinery brings the opportunity to get even bigger and realize a world of new capabilities, as our ant would have experienced when starting its hypothetical workday.

Bodies may make a dramatic appearance in the fossil record, but if the genomes of living creatures hold any clues, changes were under way for a long time. The first 2.5 billion years of the history of the planet were entirely devoid of big creatures; then, by about 1 billion years ago, there were not one but several different species with bodies populating the ancient seas: plant bodies, fungal bodies, and animal bodies, among others. The origin of bodies wasn't some magical event. The molecular tool kit that makes bodies and their organ systems possible—the proteins, large lipids, and other large molecules that allow cells to stick together and communicate—is not unique to creatures with bodies. Antecedents are present in small single-celled creatures that use versions of these same molecules to feed, to move about, even to communicate with one another. The biological mechanisms needed to build big creatures existed on the planet for billions of years before those creatures ever hit the scene.

What opened the floodgates and turned this potential for big creatures into a tangible reality? Insights, yet again, come from iron-ore-containing rocks.

Standing a wiry five feet six inches tall, Preston Ercelle Cloud Jr. (1912–1991) was one of the most imposing figures in all of pale-ontology in the postwar decades. Graduating from high school with a lust for travel and the outdoors, he entered the navy for three years, where he became the bantamweight boxing cham-pion of the Pacific Scouting Force. Cloud put himself through school during the Great Depression, eventually rising through the academic ranks to become chief paleontologist for the U.S. Geological Survey. He was a stickler for detail in his fieldwork and commanded respect from his staff. When mapping rocks, he would often crawl on them, putting his eyes inches from the layers. During one of these sessions, he was slithering through juniper thickets in Texas and came face-to-face with a large rat-tlesnake. As his field colleague at the time said, "Pres was not easily bluffed: after a few minutes of staring at each other, the rattlesnake crawled away."

Cloud had a talent for using close-up encounters with rock layers to see the global picture. In his eyes, the planet was one big interlinked system: the history of life and the workings of climate, oceans, and continents formed a single unified narrative. If the search for iron led to the discovery of early living creatures, then the iron itself led to understanding their links to the planet.

Preston Cloud.

Iron-rich layers begin to appear in rocks about 2 billion years old on every continent. No matter whether on Australia, North America, or Africa, they generally form the same series of precisely layered reddish-brown bands. As anyone who has left wet tools in the garage knows, the color is a clue to the iron's chemistry. Oxygen in the air turns iron reddish brown as it rusts.

This kind of rust is absent from the oldest rocks in the planet. When Earth formed over 4.5 billion years ago, the only major source of atmospheric gases was Earth itself. Volcanoes spew all kinds of molecules but precious little oxygen. We'd have an easier time breathing on the top of Mount Everest than on this ancient Earth. The bands of rust in more recent rocks reveal the change: a global increase of oxygen in the atmosphere.

The oxygen in the atmosphere exists in a balance between the entities that produce it and those that consume it. Like a bathtub faucet running with a partially open drain, the level in the tub is the outcome of the rates of inflow and outflow.

Clues to the inflow of oxygen in the ancient atmosphere come from ancient life itself. Most of the single-celled creatures found in the oxygen-poor world had one very important thing in common. Judging by their closest living relatives, they used photosynthesis to make energy from the chemicals around them. Photosynthetic creatures use energy from the sun to make usable energy for their bodies. They do not use oxygen; they produce it. The only possible source for the oxygen that changed Earth

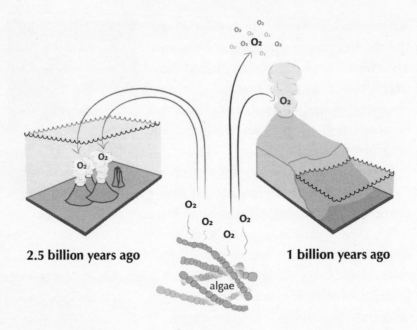

2.5 billion years ago   **1 billion years ago**

algae

Oxygen exists in a balance between the forces that produce it (algae) and those
that consume it (reactions with rock, water, and gases).

is one that we see today—photosynthetic creatures. Today that
bunch of organisms includes microbes and plants, but billions
of years ago there was one main suspect. We see it inside the
rocks that Tyler and Barghoorn peered at under their micro-
scopes. With cell walls and distinctive colonies that take a range
of forms—small clumps to toadstool-shaped masses—they look
much like modern algae. Algae, quietly producing oxygen for
hundreds of millions of years, gave breath to life on Earth.

If algae were the source of oxygen in the ancient atmosphere,
what happened to the sinks that served to remove it? Some molecules in the atmosphere can remove free oxygen from the air by
binding with it to make new compounds. Volcanoes under the
sea, for example, belch gases that are derived from the melting of
the crust at the ocean bottom. These gases have a special feature:
the molecules inside bind to oxygen, removing it from the air.

There are a number of ideas to explain the rise of oxygen in the atmosphere. One hypothesis is that around 2 billion years ago, when oxygen became prominent, there was great geographic change: a reduction in the amount of oceans and the number of undersea volcanoes emitting gases that would remove oxygen from the air. With algae pumping oxygen and fewer mechanisms to consume it, the oxygen in the atmosphere increased.

Cloud put the different observations—bands of iron, algae pumping oxygen, and the origin of big creatures—all together. The inspiration for his synthesis lies in the structure of the oxygen atom itself. Oxygen is an atom that is greedy for electrons because it lacks two of them in its outer shell. The powerhouses of our cells—mitochondria—along with aerobic bacteria make use of this fact in their energy processing. Some metabolic reactions, such as respiration, have elaborate cascades that transfer electrons from one molecule to the next, where, at each electron-transfer step, energy is either stored in new forms or released. The more free oxygen there is about, the more fuel there is for living creatures to use.

Cloud also knew that being big costs energy. The proteins, such as collagen, that form much of our bodies take a relatively large amount of energy to build and maintain. The growth and maintenance of a body require a new kind of higher-level energetics.

Clues to the relevance of oxygen levels to animal life came from the observations made as early as 1919 by one of the fathers of the field of physiology. Schack August Steenberg Krogh (1874–1949) was fascinated by physical and chemical laws that govern the physiology of animals. One of them he found by looking at the correlations between the properties of water and the creatures that live in it. In simple marine creatures, ones with no elaborate circulatory or digestive systems, body size is limited by the amount of oxygen in the water. In oxygen-restricted environments, being big is physiologically impossible, so these creatures are restricted to small sizes. With enhanced levels of oxygen comes an increase in the sizes that creatures can attain.

Preston Cloud inferred that rising levels of oxygen—and the new energetic landscape it brought—defined a new realm of possibilities for ancient life. Oxygen lifted the lid on big: it fueled the push from microscopic creatures living in a world dominated by intermolecular forces to a planet with ever-bigger species with new kinds of bodies.

Oxygen also created an entire new world of danger.

Every change is a double-edged sword. The chemistry that makes oxygen so efficient at generating energy can turn it into a poison. A great receptor for electrons from other atoms, oxygen generates energy while it forms new compounds. Unchecked, these molecules can disrupt cells and damage DNA. A number of theories of disease and aging are based on these properties of oxygen. Every time you take an antioxidant like vitamin C, you are trying to fight the effects of these kinds of oxygen-containing molecules.

But there is a deeper challenge to being a big creature living in an oxygen-rich world. A human body is composed of an enormous number of parts—two trillion cells, thirty thousand genes—that function as a whole: organs, tissues, and genes work together to ensure the integrity of each individual. The balance among parts is defined by the ways our different cells attach to each other, communicate, and interact with the molecules that lie between them. When we're healthy, each organ of our bodies "knows" how to behave; its cells continually divide and die, yet it remains roughly the same size and shape. Each eye is about the same size, as are each thumb and big toe. Our spleen and liver have their right sizes. This intricate harmony of parts is necessary for us to function as many-celled individuals.

Insights into this balance came from odd flies: members of a laboratory at Johns Hopkins noticed some flies in their breeding colony that had eyes way too large for their heads—five times bigger than normal. Being geneticists, they were well prepared for analyzing the genes. They isolated the DNA of the gene, traced its molecular activity, and discovered that it lies in a chain of molecular reactions that tell cells to stop growing. Controlling growth, either through slowing down the division of cells or by managing their death, is the centerpiece of building a harmonious large body.

Using their knowledge of the structure of the fly gene, the Hopkins team was able to trace it to mice and people. Not only is a version of the fly gene present in the genome of mammals, but mutations bring about organs of new sizes. The gene appears to behave in generally similar ways in flies, mice, and people, working as part of a molecular chain reaction to define a balance among parts. In mice, a mutation can bring about a liver five times the normal size.

But mutations of the gene also have another consequence. When cells lose their cues to stop dividing or dying at the right time, they can form deadly malignancies. The genes that define the ways our bodies get big also form the basis to destroy them.

Preston Cloud and others who followed him saw the harmony between living systems and planetary ones. The interplay between living things and their planet led to increasing levels of oxygen in the atmosphere. Oxygen, in turn, changed the world by allowing for the origin of big creatures with many cells. Life changes Earth, Earth changes life, and those of us walking the planet today carry the consequences within.

CHAPTER SIX

CONNECTING THE DOTS

200 million

E arth of 530 million years ago was still far from being a place we'd recognize as home. Breathing would have been like summiting Mount Everest without an oxygen tank; atmospheric oxygen levels, while having increased substantially since the early days of the planet, remained only a fraction of those today. Water was the happening place for life: new kinds of soft-bodied animals swam in the oceans. Land, by contrast, was a barren void lacking any sort of plant cover or soil. If you managed a hike on this strange landscape, you could have walked from what is today Boston to Australia without ever seeing ocean.

The modern world came about through changes to rock,

air, water, and life. But, as the story of oxygen reveals, these constituents do not exist in isolation from one another. Earth's history is the product of interlinked changes to the planet and the creatures on it. Learning to see the deep meaning of these grand connections, and the roots for one of the greatest scientific revolutions of all time, begins with one of the most seemingly mundane exercises of all—making maps. Maps tell us where we are and what our world looks like, and, when we use them in the fullness of geological time, expose the links that exist among oceans, mountains, and the organs inside our own bodies.

Children, as every parent knows, are driven to find patterns in the world around them. Take a globe or a map, and show it to kids about seven years old. Ask them to describe what they see, not the names of the regions, but the shapes of various continents, islands, and oceans. Almost invariably, kids will point out that parts look as if they fit together. The east coast of South America and the west coast of Africa seem to match perfectly.

In some quarters, even as recently as forty years ago, this simple observation reflected complete heresy. In junior high school, I had a teacher who lectured class on the pseudoscientific ideas that were in vogue in the late 1960s. One was the notion that the Egyptian pyramids, the Nazca lines of Peru, and the statues on Easter Island were influenced by the work of space aliens who would regularly visit our planet. Like most of the students, I thought the space alien idea great—it was on TV, after all. My teacher's intent was to show that while these kinds of ideas may have been fun and fantastic stories, they could not be tested in any meaningful way. He lumped these tales of science fiction with another called continental drift, the idea that the continents have moved over time. This lecture transpired in 1972. Little did my teacher know of a great revolution that had its start nearly 120 years before.

In 1856, the brothers William and Henry Blanford arrived in

India to work in the famed Talcher Coalfield. This field is still producing today, as part of Mahanadi Coalfields Limited. While studying the layers of coal, the Blanfords, like all good geologists, looked above and below to see geological events surrounding its formation. One of the layers stood out; it was filled with huge, irregular boulders—some even the size of a person. The mystery deepened when they looked closely at the buried boulders themselves: each had slashes and gashes on the surface. If found in the Alps or in the high Arctic, these features would have been taken as evidence of transport by huge glaciers. But here, buried underneath layers of coal in equatorial India?

The boulder layer didn't stop in India. Soon after the Blanfords' discovery, geologists working in South Africa discovered the same series of rocks: a coal bed and a layer of irregular and gashed boulders. Were there glaciers in South Africa and India at some point in the distant past? How did ice get to these places?

Eduard Suess (1831–1914) was born in London and, after moving to Vienna, developed a passion to understand the workings of Earth. After school, Suess rose through the ranks to become a leading academic and, later, city councilman for Vienna. There, he put his knowledge of geology to use for the common good by spearheading the development of an aqueduct to bring freshwater into the city from the surrounding mountains. This act is thought to have saved a large number of lives, since typhoid outbreaks from dirty water were common in the city beforehand.

Suess's view of the importance of rocks was revealed in a speech he gave to an international congress of geology in 1903, when he described the work of a geologist in this way: "The stone which his hammer strikes is but the nearest lying piece of the planet, . . . the history of this stone is a fragment of the history of the planet, and . . . the history of the planet itself is only a very small part of the history of the great, wonderful and ever changing Kosmos." To Suess, each rock, properly interpreted, was a window into a world that stretches across the far reaches of space and time.

With this philosophy as a guide, rocks and fossils took on new meaning. About a decade before the aqueduct proposal, Suess became interested in an enigmatic fossil with leaves shaped something like a cow's tongue. The plant, known as *Glossopteris* (Latin for "tongue leaves"), was a real mystery. As is typical with larger fossil plants, the whole organism is almost never found; individual leaves, branches, roots, and trunks are used to reconstruct it. This exercise is like putting a three-dimensional puzzle together from incomplete parts. By Suess's time, *Glossopteris* was already known to be strange; it had a soft woody interior, making it like a conifer or fern, but it was unlike either in that it carried organs with seeds inside. Judging from the relative sizes of branches and leaves, *Glossopteris* occasionally rose ninety feet high and may have tapered toward the top, much like a Christmas tree.

Suess noticed that the really astonishing fact of *Glossopteris* lay not in its strange leaves but in the rocks that held it. As he followed rock layers with *Glossopteris* inside, he was able to trace them from South Africa to India all the way to Australia and South America. To Suess, the distribution of *Glossopteris* meant one thing: these distinct continents were once connected to each other in the distant past. In his thinking, the continents were continuous until the seas rose to separate them.

Glossopteris figured in both discovery and tragedy. In 1912, Robert Falcon Scott, with four crew members, made a fatal attempt on the South Pole, only to find, upon arrival, that the Norwegian Roald Amundsen had beaten him by several months. Photographs reveal the group's predicament. They were clearly weakened by the long trek, and their faces in the foreground tell of their fatigue and disappointment while standing against a background of Norwegian tents and flag. Scott's diary records the difficulties the team experienced pulling heavy sledges on the return trip, how with steadily weakening bodies the weight became too much to bear. Scott, Henry Bowers, and Edward Wilson died in their tent in March 1912, and their bodies were recovered when winter broke eight months later. Lying next to their bodies were thirty-five pounds of rock and fossil. When the samples were brought to the experts at the British Museum, their significance was revealed. The team had discovered *Glossopteris* at the base of Beardmore Glacier, a hundred miles from the South Pole. Suess didn't know of this, but it meant that Antarctica too was part of the connections he was envisioning for southern continents.

The deep meaning of boulders, fossils, and the jigsaw pattern of continents emerged from the mind of an iconoclastic German meteorologist. Alfred Wegener had two major scientific passions in life: understanding the weather over the ice sheets of Greenland and the geography of Earth. He began his career in

Glossopteris, the reconstructed plant and a fossil leaf.

1911 serving as a member of some of the earliest scientific explorations of Greenland, including one venture crossing the entire ice cap on foot. His life ended on the island, where on an expedition in 1930 he died on the ice sheet in an effort to rescue some of his crew who were in need of relief.

Alfred Wegener in his element.

Wegener proposed that the different continents were originally connected as one huge supercontinent in the distant past. Over time, the continents drifted apart, forming the configurations of the oceans and coastlines we see today. The first split would have been between a northern chunk and a southern one. This southern one contained what is today Africa, South America, Australia, India, and Antarctica.

This simple idea answers many questions. Why the special similarity of the plants from the southern continents? Because they were originally part of the same landmass, after the supercontinent broke into chunks. Why glaciers at the equator in India? Because India wasn't always at the equator; it drifted over time from a position closer to the poles. And why the jigsaw shape of the continents? Because they all fit together at some point millions of years ago.

What about the response to this grand unifying concept of Wegener's? Familiar with *Glossopteris* and the geological similarities among Africa, India, and South America, geologists working in southern continents and Europe tended to view the idea favorably. The reception in North America was something else altogether. The comments of one of my predecessors here at the University of Chicago, in 1920, summarizes the state of affairs: "Wegener's hypothesis in general is of the footloose type, in that it takes considerable liberty with our globe, and is less bound by restrictions or tied down by awkward, ugly facts than most of its rival theories."

Wegener's critics agreed that the coastlines might have looked like matching jigsaw pieces, but to them the similarity was more coincidence than reality. They could envision no engine that could move the continents. Did the landmasses roar through the ocean crust like icebreakers through pack ice, crunching and crushing miles of rock along the way? Nothing in science spoke to this. In fact, everything we knew of the seafloor at the time spoke to the opposite: the bottom of the ocean appeared to be one of the calmest and most featureless places on Earth.

Of course, most of the planet was a total mystery in the early part of the twentieth century. Close to 70 percent of Earth is covered by ocean, and in Wegener's day we knew more of the bright surface of the moon than of Earth.

GREAT DEPTHS

December 7, 1941, the day of infamy at Pearl Harbor, had a major impact on our understanding of the planet. Harry Hess, a young geologist at Princeton, was called to war and, being a naval reservist, shipped out from Princeton to New York City to report for active duty on December 8. When he arrived at the headquarters on Church Street, he was asked if he "knew about latitude and longitude." Little did Hess's recruiters realize that years before he had been at sea on expeditions to explore and map features of the ocean floor. Hess's answer to the question was likely satisfactory, as he rose to become navigation and executive officer on the USS *Cape Johnson,* a maritime freighter converted into a troop transport. The *Cape Johnson* went to the South Pacific and served during battles at Guam and Iwo Jima. Hess, ever the geologist, had an additional mission in mind during these battles.

On board the *Cape Johnson* was a device known as a Fathometer, a simple kind of sonar that measures the depth of the ocean. Modern versions of these devices are small enough to carry on a canoe—think Fishfinder—but during World War II these were the size of a small refrigerator and were towed behind the boat. Hess found a painless way to do science during wartime: just leave the Fathometer running while the *Cape Johnson* performed its military duties.

Harry Hess on duty.

This effort didn't cost Uncle Sam much, but it had a large impact on Hess's thinking. Hess found a number of small flat-topped mountains on the seafloor. These little submarine mesas were to have a major impact on science fifteen years after the war—insights that were kindled when he heard of the work of another person whose life's trajectory was changed by Pearl Harbor.

Marie Tharp had no inkling of a career in geology when she matriculated at Ohio University in Athens in the late 1930s. Her path was toward a "proper" woman's career at the time: nursing or teaching. Nothing clicked. Scared of blood, she left nursing to take courses in education. That curriculum didn't excite her either. Her opportunity came after the call-up on December 7, when millions of men were pulled from their jobs to fight the Axis powers. The Department of Geology at Michigan broke with long tradition and offered new opportunities for female students to study. The geology scholarship seemed as good a gig as any, so off Tharp went to study Earth.

By the late 1940s, with her earth science degree in hand, Tharp had gone to New York City to find a job. Her first stop,

in the paleontology department at the American Museum of
Natural History, was not promising. When she inquired about a
position preparing fossils for research and display, her blood ran
cold when a paleontologist told her that it took up to two years
to discover and remove fossils inside rocks. Tharp later said
that she "couldn't imagine devoting so much time to something
like that." Paleontology's loss was geology's gain. Tharp went
to Columbia University to meet the head of a major geological
team there, Maurice "Doc" Ewing.

Doc Ewing was a big Texan who was leading an effort to map
the seas. In the shift from World War II to the Cold War, one
thing remained constant: submariners needed to know about the
structure of the ocean floor. The Office of Naval Research filled
the demand by funding expeditions worldwide to look at the
ocean's depth and structure. Ewing was sending boats throughout the year to collect cores, depth readings, and other information from the ocean bottoms. With so much data coming in, he
required someone to compile and map them. Tharp was hired
first; later Ewing recruited an Iowan named Bruce Heezen to
direct her and the mapping effort. Heezen swiftly rose through
the ranks to become a professor at Columbia, while Tharp
remained as his assistant.

These were heady days in geology. Each month, Doc Ewing's
boats returned with reams of new data from completely unexplored regions of our own world. Tharp and Heezen were in
the middle of this frenzy, synthesizing much of the data Ewing's
teams collected. The two worked hard and became very close,
but in a way nobody could entirely comprehend. Heezen, a mar-
ried man, often entertained students and colleagues at Tharp's
house near the laboratory. Some days they would battle, with
Heezen throwing Tharp's drawings in the trash or screaming
insults that shattered the normal quiet of the halls. Other times,
the two would behave as a team, defending each other during
vicious political battles that were part of the environment of the

Bruce Heezen and Marie Tharp.

lab. Always close, Tharp and Heezen had a relationship that was, by all accounts, emotionally intense but entirely platonic.

One day, after countless hours compiling shipborne records of the ocean depth, Tharp saw a linear chain of mountains over a mile high that extended along the floor of the Atlantic Ocean. There were solid hints before that these ridges existed, but she followed them as they coursed forty thousand miles on the bottom of the ocean, through virtually every ocean on the globe. Then she looked at the structure of the ridges themselves. Within the apex of each ridge sat what looked like a giant valley— a depression that split the ridge in two. The walls on either side of this valley appeared to match up. Tharp had a hunch what this implied: Earth was opening up at the ridges as it rifted apart at the bottom of the ocean. To her, this was evidence that the seafloor was expanding. Excited, she approached Heezen with the idea.

Heezen hated Tharp's discovery, calling it "girl talk." Like Tharp, he saw the implication immediately. To him, Tharp's rift in the center of the seafloor looked "too much like continental drift." If the middles of the oceans were separating, then the continents were moving apart, and Wegener would be right after all. Heezen couldn't abide this speculation.

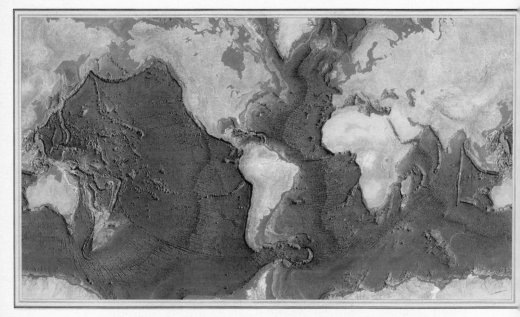

Tharp and Heezen's map with the giant ridges in the center of the ocean.

But Tharp's data did not go away; in fact, the more she plotted, the more her rift became obvious. Heezen's resistance withered with the mountain of new data that emerged over the ensuing months. He not only became sold on Tharp's idea, but he came up with the even more ambitious plan to make a map of the entire ocean floor.

Around this time, American Telephone and Telegraph realized it had a problem with transatlantic cables that were breaking frequently. The company contracted with Ewing's lab to check the situation. Plotting the seismic data Ewing's team collected at a fine scale, Heezen, Tharp, and the team found a stunning pattern. The earthquakes ran in a regular line in the ocean. And not just anywhere in the ocean; they did so in the middle of Tharp's rift valleys. People started to become very interested in Tharp's rifts.

"Girl talk" became the subject of a professional seminar Heezen gave to the assembled experts in the geology department of Princeton University in 1957. In the crowd was Harry Hess, now chairman of the department. After seeing Heezen's presentation of Tharp's rifts and their earthquakes inside, Hess rose to say, "Young man, you have shaken the foundations of geology."

Hess was primed to love Heezen's talk because of his work during World War II mapping submarine mountains: they revealed a pattern similar to those of Tharp's ridges. His mountains were high near the big ridges and became eroded the farther away he looked. To Hess, this meant that the mountains closer to the ridges were relatively young; those farther from the ridges, old. Along with the data revealing active splitting at the ridges, the only explanation could be that new seafloor was created at the ridge and the seafloor was indeed spreading.

Geological work at this time was an international effort filled with story after story of discovery and persistence. One six-foot-five-inch Dutchman lay curled up in a tiny submarine for weeks on end while mapping deep-sea trenches. British, Canadian, French, Dutch, and Japanese scientists spent months on board ships mapping coasts, oceanic ridges, and trenches. All of this activity brought the need for a new view of Earth. With data pouring in from around the globe, the deep-sea trenches started to reveal a pattern: they too were the sites of frequent earthquakes and, on many occasions, volcanic magma emerging at the surface.

Heezen's presentation stimulated Hess to devise a theory to explain the different observations. If new seafloor is created at the ridges, then it had to be recycled somewhere else, lest Earth be ever expanding. To Hess, the pattern of earthquakes and other physical features of the deep trenches fit the bill. He proposed the notion that seafloor emerges at the ridges, spreads as it moves away, and later sinks and is destroyed at the trenches. The seafloor is now seen to be a huge conveyor belt.

Hess wrote up this idea in a manuscript that he circulated among colleagues but hesitated to publish for two years. He called

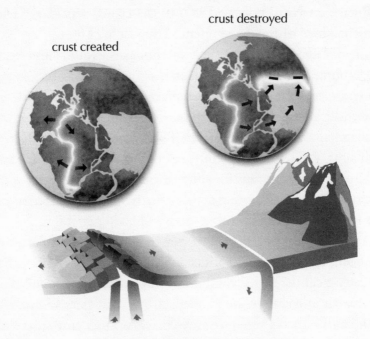

crust destroyed

crust created

Geopoetry and the recycling of ocean crust.

his idea "an exercise in geopoetry," as much to defer criticism that it was speculative as to celebrate its beauty. In fact, elements of Hess's idea, like many ideas in science, had been proposed by somebody else before. Arthur Holmes, a brilliant British geologist, derived a similar recycling idea from pure theory in 1929. Holmes, one of the pioneers in the development of modern methods of dating rocks, found his inspiration in Wegener himself.

Lacking for geopoetry were insights into the age of the seafloor; eroded mountains and rifts alone weren't going to put an end to almost a century of skepticism. Hess presented his theory at Cambridge in the early 1960s, and in the audience was a young student by the name of Frederick Vine. Excited by Hess's theory, Vine and his adviser, Drummond Matthews, hunted for an indicator of the age of the rocks at the bottom of the ocean so that they could compare the age of the seafloor on either side of Tharp's rifts. The two developed a clever technique using the data they had at hand. If the seafloor was spreading like a conveyor belt, then the youngest seafloor should be close to the ridges, and the ages should increase as you move away. Furthermore, the ocean floor should be the same age at the same distance on either side of the ridge. Vine and Matthews used the pattern of magnetism inside the rocks of the seafloor as a marker for their ages. They found exactly what was predicted: young floor is close to the ridge, older floor farther away, and the ages on either side of the ridge match. The seafloor was spreading, just as Hess and Holmes before him had proposed.

At the same time Vine and Matthews were preparing their publication, Lawrence Morley of the Canadian Geological Survey was assembling his own data. He submitted his analysis to the august journal *Nature*. It was rejected. He then submitted it to the more specialized *Journal of Geophysical Research* in 1963. Several months passed. Then it was returned with an anonymous note from one of the referees saying, "Found your note with Morley's paper on my return from the field. His idea is an interesting one—I suppose—but it seems most appropriate over martinis, say, than in the *Journal of Geophysical Research*." This delay cost Morley dearly; soon after he got news of his rejection, Vine and Matthews's paper appeared.

Vine and Matthews did not measure the age of the seafloor directly; the technique they used was so new that it required refinement before Hess's geopoetry was to become universally accepted. Confirmation came only a few years later with more surveys of the ocean floor led by Columbia, Stanford, and the Scripps Research Institute in La Jolla, California. With the mountains of data, new ideas, and Wegener's classic insights, *Time* magazine produced an article in 1970 with a title that says it all: "Geopoetry Becomes Geofact."

For professors like Hess and Heezen, this revolution in thinking led to fame and academic eminence. But old feuds die hard. Because of spats with Ewing, largely due to their support of continental drift, Heezen and Tharp had become persona non grata at Columbia. Heezen, a tenured professor, could not be fired, but even so Ewing found ways to demean him: he stripped Heezen of his departmental responsibilities, cored the locks from his office door, dumped his belongings in the hall, and gave his office away. Ewing did manage to fire Tharp. Lacking an office, she ended her career working out of her Nyack, New York, home. Her view of the tumultuous personal and scientific times was revealed twenty years after Heezen's death when, during an oral history project at Columbia, she recalled, "I worked in the background for most of my career as a scientist, but I have absolutely no resentments. I thought I was lucky to have a job that was so interesting. Establishing the rift valley and the mid-ocean ridge that went all the way around the world for 40,000 miles—that was something important. You could only do that once. You can't find anything bigger than that, at least on this planet."

ALL TORN UP

If Wegener's continental drift evokes gradual movement and Hess's geopoetry a sublime relationship between parts of Earth, then the word that describes the merger of the two, "tectonics," conjures an idea that rattles our world to its foundations. The 1960s had revolution in the air in music and politics, but arguably the most lasting change was the emergence of a new way of seeing the planet. Patterns of rocks and fossils that were once bizarre started to make total sense. Scientists were gleefully revising centuries of scientific dogma, and foremost among these revolutionaries was John Tuzo Wilson, a Canadian geologist. Wilson, a physicist by training, had a personal quality that was to serve him well during this time. Late in his career

John Tuzo Wilson, when he wasn't
disturbing scientists.

he summed it up: "I enjoy, and have always enjoyed, disturbing scientists."

In the late nineteenth century, paleontologists recognized that North America and Europe both have distinctive mountain ranges that run north to south. The Appalachian Mountains extend from Maine to North Carolina in the United States, and the Caledonians course from Morocco to Scotland in Africa and Europe. One place in northern Scotland has Appalachian fossils inside its rocks. And in a few places Appalachian rocks have European fossils inside. How did the creatures get there? One could invoke swimming, except that most of these shelly animals live fixed to the seabed and do not travel far.

To Wilson, armed with the new theory, the answer to this puzzle was akin to what happens when a child plays with a peanut butter and jelly sandwich. What happens when the child takes the two separate halves of the sandwich, puts them together, and then opens them up again? The mushed jelly and peanut butter reflect what happened when the sandwich closed up. The bits of peanut butter left on the jelly side and jelly on the peanut butter side reflect the opening.

To Wilson, Europe, North America, and the Atlantic Ocean behaved the same way as that sandwich. He proposed that there was an ocean that separated the continents hundreds of millions of years ago. This ocean closed, and as the continents from either side rammed into each other, a chain of mountains formed. When the continent reopened, the chain broke into two pieces, leaving what are today the Appalachians and the Caledonians. And those odd American fossils in Europe? They were just patches of the old continent that got left behind when the Atlantic Ocean opened.

Wilson, and the many scientists who followed, found the globe that we learn in school is only a snapshot in time: there have been innumerable globes in our past, and there will be many more in the future. Earth's crust is composed of a number of plates, each containing ocean, continent, or both. These plates move relative to one another as the convection under the crust causes the seafloor to spread at Tharp's ridges and to ultimately be destroyed at the deep ocean trenches.

In 1984, over half a century after Wegener's death, NASA released the first direct measurements of continental drift. About twenty stations around the world were established, each capable of bouncing lasers off satellites equipped with reflectors. A telescope next to the laser on the ground picked up the reflection by the satellite. By measuring the time that laser light took for the round-trip to each station, NASA calculated the distance to the satellite. If the plates move, then the distance to the satellite should change over time. Using this technique, NASA showed that North America and Europe are getting farther apart by 1.5 centimeters per year. Australia is heading for Hawaii at about

Connecting the dots between the rocks and rifts in North America and Africa.
The shaded areas hold matching rocks.

7 centimeters per year. The plates on our planet move about as
fast as hair grows on our scalps.

The motion of the plates is slow over the pace of our life-
times, but over geological time it can be cataclysmic. This dance
of the plates takes several major steps. Plates can move against
one another. As they rub, the plates can experience earthquakes,

like those at the famous San Andreas Fault in California. Some plates smash into each other. When the plates are continents, this collision results in new mountain ranges. The Tibetan Plateau came about when India started colliding with Asia over 40 million years ago. Plates can also move apart. If, for some reason, upwelling of the convection current under the earth happens in the ocean, we see Heezen and Tharp's ridges and rifts. When this upwelling happens under a continent, this single patch of land can rift into several.

Plate tectonics reveals connections everywhere. Farish and the team were led to Greenland, as we saw in the first chapter, by similarities to the rocks we knew from our work in Connecticut and Canada. In each location ancient faults, lake sediments, and sandstones point to one event in eastern North America 200 million years ago: an opening rift in the earth. Those similarities extend all the way across the Atlantic Ocean to the rocks in Morocco and Europe. The pattern of the rocks and fossils allows us to connect the dots: at one time eastern North America, Greenland, and Morocco were the same continent, which then broke apart during the formation of the Atlantic Ocean 200 million years ago.

Maps and rifts, however, link more than the features of the globe.

THE MAP WITHIN

In 1967, the Levingston Shipbuilding Company of Texas laid the keel for the *Glomar Challenger*, a ship that looked like any other save a giant drilling tower that rose nearly fifty feet from its center. For the next fifteen years the *Glomar Challenger* traveled the seas drilling cores in the seafloor. Making more than six hundred stops, it was able to dredge up cores of rock from almost two thousand feet beneath the bottom of the ocean. Each long core would come to the surface as fifty or more thirty-foot-long strips of rock and sediment looking something like gray-brown flagpoles. These cores, almost twenty thousand of them in all, became a bonanza for science: they told us the age of the seafloor, its composition, and its history. Today, they lie in repositories around the world, where they are still being studied long after the *Glomar Challenger* was scrapped.

Locked inside the cores are layers of minerals with combinations of atoms that allow us to reconstruct the atmosphere, temperature, and workings of the planet for the past 200 million years. Looming large among these are atoms that reflect the levels of oxygen in the atmosphere. Parts of the analysis might seem counterintuitive, but the central idea is that oxygen concentrations in the atmosphere can be approximated by measuring carbon in its different forms. Carbon and oxygen exist in a balance on the planet: carbon ejected from volcanoes interacts with and influences the levels of oxygen in the air and water. By measuring the different atoms of carbon in any given layer of sediment in a core, we can approximate the levels of oxygen in the air.

Two worlds lie inside the cores drilled by the *Glomar Challenger:* one from before the Atlantic Ocean started opening 200 million years ago and one that emerged after. The oxygen in the atmosphere increased dramatically after the Atlantic formed. By about 40 million years ago, the atmosphere went from one in which we would pant merely to sit still to the one we run around in today.

The rift that began to open over 200 million years ago and split the supercontinent into multiple bits created enormous amounts of new coastline. Each coast is an area where land meets the sea. As every coastal homeowner knows, these areas are subject to erosion. A dramatic increase in erosion can set off a chain reaction. Imagine entirely new coastlines dumping sediment into the sea. With this sediment comes the burial of very special mud that covers the bottom of the ocean shelf. This

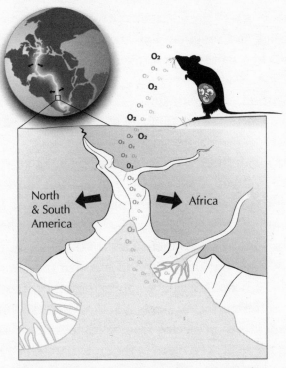

The opening of rifts and the chain reaction that
enhanced oxygen in the atmosphere provided opportunities for our ancestors.

muck is extremely important, because every day trillions of
single-celled creatures die and sink to the bottom; as they decay,
they consume oxygen. Left alone, this mass of waste eats enormous amounts of oxygen from the water—and ultimately from
the atmosphere—as it rots. But when these layers get buried,
oxygen is no longer consumed as quickly, allowing it to build
up in the water and the air. This is what the rifts and new coastlines have wrought: increasing levels of oxygen in the air brought
about by the burial of oxygen-consuming muds.

A new world began with the rift, one with ever-rising levels
of oxygen. And, as we've seen, with oxygen come opportunities.

Mammals like us are committed to a very high-energy lifestyle. We manufacture our own heat. The action of our muscles,

coupled with the insulation provided by our hair, fat, and, in the case of humans, clothes, keeps our body temperature stable relative to the outside world. Cold-blooded creatures, like lizards, also can keep their body temperatures relatively stable, but they use mostly behavioral mechanisms to do this: sunning on rocks or hiding in shade. In the cold, a lizard cannot be active. I don't need to worry about snakes on my expeditions north; polar bears are our major concern. Mammals can remain active in climates that would kill cold-blooded animals. Our warm blood disconnects us from the vagaries of the temperature of the outside world. Fueling these fires requires oxygen.

Not only are we insulated from the outside world as adults; we begin our lives inside a womb surrounded by membranes that protect the embryo and provide it with connections to the mother's blood supply. Since the fetus receives all of its oxygen from the mother, there needs to be a way that oxygen can be transferred from the mother's blood. The transfer is facilitated by a steep gradient between the concentration of oxygen in the maternal blood and that of the fetus: under these conditions, oxygen will travel into the fetus. Importantly, the oxygen content of the mother's blood has to be sufficiently high to enable this transfer in the first place. This constraint means that mammals with a placenta do not easily develop above fifteen thousand feet altitude. Tellingly, the oxygen at these altitudes is equivalent to that in the atmosphere at sea level 200 million years ago, before the Atlantic Ocean formed.

Insulation of bodies from changes to the outside world comes at a cost: a big warm-blooded mammal needs fuel to maintain a constant body temperature, develop in the womb, and thrive outside it. Oxygen is the key link: animals like these could never have emerged in the low-oxygen world that existed before the continents split apart. Marie Tharp's rift didn't only open up an ocean; it opened up a whole new world of possibilities for our ancestors.

CHAPTER SEVEN

KINGS OF THE HILL

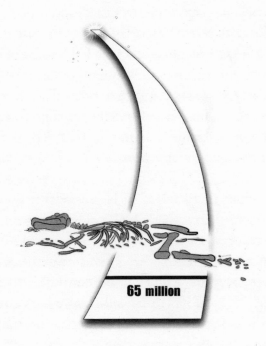

65 million

J ust one more step," Paul Olsen kept repeating like a mantra.
He was urging me on, but I was frozen like a cat in a tree.

We were on the shores of Nova Scotia taking a break from
fossils to collect geological samples. The coastline is made of
spectacular red, orange, and brown sandstones that are remi-
niscent of the Hopi and Navajo reservations of the deserts of
the American Southwest. The beauty of this place is magnified
by water: rocky bluffs erode into a natural sculpture garden of
caves, arches, and pillars. Paul, a geologist at Columbia Univer-

sity, wanted to obtain sand grains from a white layer that separated orange rocks below from brown ones on top.

Unfortunately, this band of white rock was about two hundred feet up a sandstone cliff that was in places almost too sheer on which to stand. In others, it was so highly weathered that one misstep could lead to a long tumble down. To get traction in these places, we had to climb step-by-step using footholds that we carved with our rock hammers. Not being a climber, and moderately scared of heights, I had made progress by only looking at my feet, hammer, and hands, knowing that even a momentary glance down the cliff could summon a rush of vertigo that would freeze me in place. On previous occasions, this panic usually brought the assistance of a team of patient colleagues who formed a human bucket brigade to coax me down to the beach below.

An hour or two of Paul's cajoling propelled me to the layer. Up close, the white band was about as tall as a human. For about an hour we chiseled rocks, placing small specimens in labeled bags for analysis back home. Our reward came when we looked at the vista of the Bay of Fundy in front of and below us. It was a glorious early summer day: the tides were high, the wind low, and the bay so smooth it looked like reflective glass. The splendor of the bay reveals its history. The shape of the coastline reflects the long-term action of glaciers, faults, and erosion. The pasture and human settlements form a recent veneer on this ancient landscape. Layer after layer of history reveals itself when you know how to look.

It was the vista inside the rocks that drew Paul's attention. The white band as well as the composition of the rocks surrounding it brought us here, because inside lie clues to events that shaped our existence.

Moving continents and changing oxygen levels gave the world a decidedly modern configuration by 200 million years

The cliff in Nova Scotia from afar (left) and up close at the white band (right). Geologists for scale.

ago, except for one major thing: for millions of years, the largest animals were not mammals, as they are today, but dinosaurs and their "reptilian" cousins—mosasaurs, plesiosaurs, crocodiles, and pterosaurs. Land, water, and air were populated by an entirely different world of creatures, all of them successful by every yardstick we can apply: there were numerous species that thrived for millions of years across wide stretches of the globe. Then they disappeared.

LOST WORLDS

In 1787, William Smith was hired to assess the financial value of the land within an estate in Somerset, England. He was never to find monetary rewards; Smith's gold was mined from something else altogether.

Smith set off to survey the rocks that lay exposed along streams, on hills, and inside coal mines. Working in one of the pits of the older mines on the estate, he noticed that the rocks that border the mine were set in layers that he could easily rec-

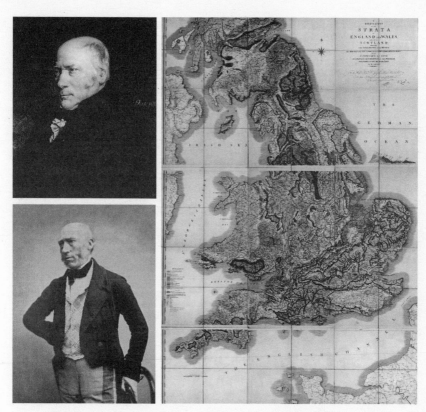

William Smith (top), John Phillips (bottom), Smith's map of England (right).

ognize by their colors and textures. On closer inspection, he discovered each layer was made of a particular kind of rock with a distinctive collection of fossils inside. Comparing these layers with others nearby, Smith had a rush of insight: the rock layers in the mine were similar to others at the surface elsewhere on the estate. As he looked closely at the layers, he saw he could use the fossils to match them in different regions, almost like a huge jigsaw puzzle. Natural philosophers, even Leonardo da Vinci, noted that it is possible to do this kind of comparison of rocks, fossils, and layers locally. Now, armed with this simple insight, William Smith had the key to map the geology of Earth: rocks and fossils arranged in layers.

Smith widened his hunt, first looking at the area around Bath, then ultimately broadening his aspirations to cover all of Britain. This new task required money, and with neither an academic post nor the auspices of any scientific society Smith was strapped for cash. He convinced about one hundred patrons to fund his effort and set off to visit every rock exposure he could. He had expert help: his nephew John Phillips had been his ward since the death of both of his parents, when Phillips was seven years old, and he accompanied his uncle on his excursions. By the age of fifteen Phillips had gained a phenomenal eye for fossils.

Today we use aerial photographs and GPS-driven survey equipment to construct geological maps, relying on comparing rocks fortuitously exposed at the surface and in deeper levels of bedrock brought up by drill cores inside Earth. This is big science, often heavily financed by oil companies, mineral interests, and governments. Geological maps are the seed corn of research on Earth: everything we do on expedition starts here. In 1815, Smith accomplished this feat largely on his own using tools of his own design. When finished, the map was a triumph. Standing seven feet high, it revealed the relative position of major layers and fossil eras throughout Britain.

Unfortunately for Smith, however, George Bellas Greenough was a leading light in the London Geological Society at the time. Without Greenough's support, Smith's map could not gain the kind of professional traction needed to sell enough copies to pay his debts. Not only did Smith fail to get Greenough's endorsement, but Greenough set off to produce his own map. And, piling the frustrations on his rival, Greenough made sure his map was cheaper than Smith's.

Smith's map was such a sales disaster that he ended up spending eleven weeks in debtors' prison, returning to find his property seized. He had hoped to keep the fossil collection he made with his nephew but had to sell it to pay his debts. The situation went from bad to worse; about this time his wife went insane and had to be institutionalized.

Despite these setbacks, Smith's legacies are many. He confirmed that the fossils in rock layers change from the deepest ones, the oldest, to the highest and youngest ones. He revealed how fossils can be used as markers to trace the same layer across a wide area. And, importantly, he gave his nephew John Phillips an eye for fossils and geological layers.

If his uncle was an antiestablishment symbol troubled by an unfortunate marriage, then Phillips was the opposite: an established Oxford don who lived with his sister for all of his adult life. Phillips devoted himself to his uncle's layers: his uncle recognized them, but Phillips was determined to find their meaning.

Work with his uncle gave Phillips the keen eye and fastidious technique that allowed him to assemble a prodigious and well-curated collection of shells, bones, and fossil impressions. Starting with his uncle's map, he traced every known fossil from every layer and asked what happens at the transition between each layer.

Phillips saw three eras of time, each with its own world of fossils inside. The differences between these lost worlds were defined by a sharp boundary where creatures simply disappeared, only to be quickly replaced by new forms of life. Phillips saw these as three major divisions of geological time and named three geological eras based on them: Paleozoic, Mesozoic, and Cenozoic. He published his findings in 1855, and if you want to know how significant his work is, just go to any museum today. You will find his three great eras plastered on the time charts adjacent to the dinosaurs, sharks, and trilobites.

This was an age of exploration of the natural world, and by looking at rocks and fossils, people began to formulate new ideas. Ships returned monthly from the far corners of the planet loaded with minerals, plants, animals, and rocks previously unknown to science. Natural philosophers of all stripes—people we would today call anatomists, paleontologists, and geologists—were in the center of the action, attempting to decipher the menagerie of biological curiosities that were being brought home.

One of the luminaries of this period, Georges Léopold Chrétien Frédéric Dagobert Cuvier, had an ego as large as his name. Born to humble origins, he died Baron Cuvier, one of the leaders of the Natural History Museum in Paris.

One expedition returned to Paris from South America with a giant six-foot-long skeleton shaped something like a small troop transport. With massive bones, large claws, and a skull with flattened teeth, this creature departed from anything in Cuvier's experience until he looked closely at the vertebrae and limb bones. Being an astute anatomist, he saw that nestled inside this bizarre skeleton is the body plan of a sloth. But it was unlike any alive today.

Then several different kinds of elephant-like bones were brought to Cuvier's attention. Seeing the differences from elephants, he identified the bones as reflecting a new species: mammoths. But these discoveries, satisfying as they were for understanding the diversity of life, raised a troubling question: Where on Earth were these creatures still alive?

Cuvier made the connection: perhaps the large sloths and mammoths were no longer roaming the planet but instead revealed lost worlds of creatures. The concept of extinction, something so fundamental to the way we see the world today and alien to many thinkers for millennia, now explained the goings-on inside the rocks.

One example after another of long-lost life appeared. Spelunkers in Germany ran across the large bones of a monster or dragon lying on a cave floor. An anatomist from the local medical school saw that they were from some sort of bear, but one so large and oddly proportioned that it was unlike anything walking

Europe. Years later, Thomas Jefferson found giant mammoths, sloths, and other creatures near his home in Virginia.

Cuvier was a big thinker, and not satisfied with mere description, he drew generalizations from his observations, putting his theories on the line. To Cuvier, the conclusion was obvious: extinction was not only real but common. So prevalent and important was this concept that in an early monograph he declared, "All of these facts . . . seem to me to prove the existence of a world previous to ours, destroyed by some kind of catastrophe."

Cuvier's idea, like that of Phillips before him, was that catastrophes shaped Earth. This idea was the cutting-edge science of its day, having the weight of evidence and the stamp of eminent authority. It also was virtually ignored by scientists for over a hundred years.

The notion of catastrophes lay in direct conflict with the reigning scientific approach of the day. This alternate notion was so powerful in explaining Earth that it did not allow interlopers. Its success was derived from the motto "Use the present to infer the past." This idea is so simple and elegant that we take it completely for granted. If you see a car parked on one side of the street on Monday but by Thursday it is on the opposite side, you infer that somebody drove it and later parked it in the new place. It would be a far stretch to imagine that the car flew by its own volition or was carried by a special wind. The mechanisms at work today explain yesterday; no magic or extraordinary physics need to be involved.

The same kind of reasoning applies to the history of rocks, cliffs, and layers. The major forces at work around us today are wind, rain, and gravity—all products of the laws of physics and chemistry. If they shape today's world, they must have acted in the past to make the rock record. The Grand Canyon clearly is a deep cut in the ground with the Colorado River at its base.

The known earthly cause for the formation of the canyon is the erosive action of water cutting the rock and the relative uplift of rocks around it. But these mechanisms are very slow. Sand doesn't compact into rock layers overnight any more than a flowing river carves a canyon thousands of feet deep in a day, or even a year. The implication for the formation of the Grand Canyon, or any geological feature, is that it took millions of years to come about.

This gradual approach provided an explanation for the formation of canyons, coral reefs, and coastlines: not only were present-day mechanisms capable of explaining Earth's history; they implied that most changes to species or the planet would be slow. Looking at Earth today, nobody could imagine, much less see, a mechanism that could bring about a global cataclysm to life on Earth.

Theories of catastrophes, like those proposed by Phillips and Cuvier, became decidedly oddball views, relegated to a kind of lunatic fringe of scientific thought. Phillips continued to work, but by the time of his passing in 1874, from a fall down the steps of All Souls College at Oxford, the notion of catastrophes was already dead—killed by the reigning dogma of gradual change.

REVOLUTION

The town of Stafford is nestled in south-central Kansas; its population consists of about a thousand households, with a high school so small they play eight-man football. In the early twentieth century, the Newells were known to locals as the go-to people for knowledge about local natural history. When farmers hit a strange rock, the patriarch of the Newell family saw it for what it really was, a mammoth tooth. Six-year-old Norman Newell observed these encounters, and they changed the way he thought about home: the flatlands that are Kansas today were once grasslands and forests that housed large mammals. Norman's interest in paleontology grew, and he excelled to the point that he won a spot in the prestigious graduate program in paleontology at Yale University, which by this time, the 1930s, had become one of the major centers of research in the field.

Norman Newell's work at Yale was a family affair; his wife supported him financially, cataloging specimens for Yale's Peabody Museum, until Newell was able to obtain a scholarship in his second year in the program. He set off studying clams, mollusks, and other animals with two shells separated by a hinge. Newell was quick to see the advantages of studying these animals. With hard shells, they readily fossilize and are very common throughout the ancient layers of the world. Newell did something few others at the time even considered important: he used living shelled animals to infer the behavior of extinct ones.

After a stint in Peru with the State Department during World War II, in 1945 Newell took a post at the American Museum of Natural History in New York City. This was a beautiful partnership: it brought Newell in contact with a renowned collection, eminent scientists on the museum's staff, and ample resources to financially support science. At this time, the museum was heaven on earth for studies of fossils and taxonomy. The area behind the scenes at the museum consists of hallways, some of which are over a quarter of a mile long. The corridors are a buzz of scientific activity. New fossils and creatures collected from around the world are coupled with new ideas about nature. It has been, and remains, a crucible for innovative ideas.

Soon after arriving in New York, Newell was asked to produce two chapters for a major compendium, the *Treatise on Invertebrate Paleontology*. The volume is as intimidating as its name. The conceit behind the *Treatise* was to produce a running compilation of every fossil ever collected, with details on its anatomy and on the layers in which it was found. This notion sprouted to what is now a vast fifty-volume set, authored by over three hundred paleontologists, each an expert on one set of fossils. To many, this kind of work seems like nothing more than stamp collecting. In the hands of some scientists, like Newell and others to follow, it is a window into a universe of scientific discovery.

Newell dwelled in the details of his fossil shells. He knew their anatomy, diversity, and, importantly, the layers of rock they came from. Like Phillips and Smith before him, he read the layers of the world as a book. Unlike them, he was now armed with vast amounts of global data, of the kind that were going into the *Treatise*.

The more time Newell and others spent compiling the fossil record, the more an inescapable fact emerged. Whole worlds of animals and plants populated the globe in the past, only to disappear rapidly and nearly simultaneously around the planet. Life had experienced not one global catastrophe but several.

Newell became a voice in a small chorus of people arguing for the reality of global cataclysms of the kind Phillips and Cuvier had argued for more than a century earlier. The response was the same: the work was largely ignored. The discovery of patterns in the past record did little to change more than a century of entrenched thinking. The theory of continental drift suffered a similar fate in some quarters: the pattern of the continents was as clear as day, but lacking any mechanism to account for moving continents, many were reluctant to accept that continents moved. The same was true for catastrophes. What kinds of mechanisms could bring about such global calamities?

In the late 1970s, Walter Alvarez, a Berkeley geologist, was working on rocks about 65 million years old in Italy. This is the time that saw the demise of the dinosaurs, a period known as the Cretaceous. Walter, an acutely perceptive field geologist, was able to pinpoint the end of the Cretaceous to a single thin layer of clay. Below this were layers of dinosaurs, marine reptiles, and other kinds of life. Above, all of these creatures were missing. Walter was asking the question: How fast did creatures die out? Answers, he believed, lay inside this clay. Perhaps a chemical inside could act as a kind of clock that he could use to estimate how fast the clay was deposited?

Walter took the problem to his dad, Luis, a Nobel laureate in physics, also at Berkeley. The elder Alvarez had a restless mind; he was always looking to apply his knowledge of particle physics to solve mysteries in science. At the time Walter approached him about his clay, his dad was thinking of ideas to scan the inside of the great pyramids for treasure.

The Alvarezes hatched plans to make refined measurements of some of the elements inside the layers. One of these was the element iridium, which is rare on Earth but common on certain kinds of asteroids and meteorites. The thinking was that if meteors bombard Earth at constant rates, iridium levels should act as a kind of clock. Iridium is found in rocks in parts per billion—the equivalent of measuring a single grain of sand on an entire beach the size of Jones Beach in Long Island. Luckily for them, the elder Alvarez was associated with a team that had the expertise, and the machines at the Lawrence Berkeley National Laboratory, to make such precise measurements.

Walter and his dad were in for a huge surprise, as iridium levels in the clay defied all of their expectations. The levels of the element were by no means regular in the layers; iridium was practically absent in most layers and off-the-charts abundant in one particular place. It was clear that asteroids didn't hit Earth at constant rates; every now and then there is the big one. And the big one they found was reflected in a huge spike in the concentration of iridium exactly at the layer that heralded one of the greatest catastrophes of all time for life on the planet.

Then Luis came up with the killing mechanism. He proposed that when an asteroid slams Earth, it vaporizes and sets off dust in the atmosphere that blocks light and kills plants. These effects cascade through the food chain, causing widespread disaster. Not only could we now imagine a mechanism for a global calamity, but we could look at the layers of rock in the world and see the effects it wrought on living things. The thrill of the scientific hunt is to have an idea whose truth is hitched to predictions that take us to new places to explore, objects to discover, and data to analyze.

The influence of the asteroid theory goes beyond rocks falling from space; it extends to how we think about catastrophes in general. For the first time in the eons that humans have looked at rocks, bodies, and fossils, not only could we imagine a mechanism for a global cataclysm, but we could reconstruct its effects and analyze its impact on the biosphere. The asteroid impact notion put catastrophism back on the intellectual agenda. The insights of scientists like Phillips, Cuvier, and Newell were no longer at the lunatic fringe of scientific thought. The question had changed from "Could catastrophes ever happen?" to "What are the consequences of global cataclysms?"

A NUMBERS GAME

In the late 1960s, Tom Schopf was a young man with a plan to transform the way we think about the past, and he didn't care if he was going to ruffle a lot of feathers in the process. As he saw it, most paleontologists worked on their own little group of animals, on their own little sliver of time. It was a field of special cases. The way we did paleontology had to change if we were to answer the really big questions. As Stephen Jay Gould once said, Schopf wanted to "rescue paleontology" by bringing numerical rigor to the discipline.

And how was Schopf going to bring this all about? Whether

he knew it or not, he was trying to bring the field back to its roots—to John Phillips.

"What can we do that's different?" With that, Schopf laid the challenge to an unusual gathering. He brought some of the leading lights in paleontology together in a conference room at Woods Hole, on Cape Cod. When they arrived, they found boxes of the *Treatise on Invertebrate Paleontology* on the table waiting for them. They were going to pick up where people like Newell left off and find new general patterns in the history of life. With some of the best brains in the field, and virtually all the known fossil discoveries yet compiled, locked for three days in a room on the shores of Massachusetts, something fantastic would happen. At the very least, this setting had the makings of an Agatha Christie mystery.

What was the result of Schopf's three-day collision between all the fossil data then compiled and some of the best brains in the field? One of Schopf's Chicago colleagues who attended the meeting summed it up: "We got nowhere. Dead zero."

Fortunately, Stephen Jay Gould brought one of his new hot-shot graduate students to the last day of the meeting. Named Jack Sepkoski, he was a computer whiz who had just graduated from Notre Dame.

There is no record of what young Sepkoski said or did at the Woods Hole meeting. After the conference, though, Gould assigned him the task of compiling the *Treatise* and other databases into a form that could be computerized by digitizing every occurrence of a fossil group on a geological timescale. This was in 1972. Sepkoski set off on tabulating things, quietly assembling the data. The job grew and grew. Sepkoski continued to crank away, even after he himself became a professor, at the University of Chicago. Ten years after the Woods Hole meeting, he unveiled the first usable database in paleontology.

By the time I was a graduate student in the 1980s, Sepkoski's database was the center of almost every debate in the field. With all the numbers crunched, it became clear that the patterns of life are most definitely not random. During the early history of animals, their diversity increases rapidly to a kind of plateau. Diversity wiggles up and down a bit, but there are five intervals where the numbers of species just crash. The most famous of these was the one that killed the dinosaurs, the so-called end-Cretaceous event at about 65 million years ago. Forever gone with the dinosaurs were the reptiles that lived in the seas, flying reptiles, ammonites, and hundreds of less-famous shelled creatures. Other extinctions happened at 375 and 200 million years ago. The general pattern looked the same for each event: species from around the world simply vanished at the same time. One of the events was nearly the end for life on Earth: 250 million years ago over 90 percent of the species living in the oceans disappeared forever.

Catastrophes were no longer pipe dreams conjured by offbeat scientists; the shape of our world was sculpted by them. And, as we've come to appreciate since the work of the Alvarezes, asteroids aren't the only likely killing mechanism. Massive volcanoes and chemical changes to the oceans have been shown to also be candidates for a number of global extinctions in which asteroids do not appear to be involved. Knowing these facts, we can now ask powerful new questions.

Who survives a global catastrophe? Are there rules that determine how life responds? Sadly, neither Sepkoski nor Schopf would live to see the progress on these big questions. Schopf was a hard-charging scientist who simply didn't have an off switch. He attacked intellectual problems and worked them around the clock. Tragically, his heart gave out during a geology field trip in 1984, stopping his work forever at age forty-four. Sepkoski died at his home in Chicago in 1999.

After Schopf's death, another young Turk, David Jablonski, was recruited to fill his post at Chicago. Jablonski's office sits across campus from mine, in a 1970s-era brick remake of a Moorish fort. Dave has a corner laboratory, a large open room overlooking the science library—or at least his room was open before his collection of thousands of books, papers, and journals filled the space. Getting to Dave's desk is a bit of a challenge. The visitor needs to meander through a maze of waist-high stacks of journal articles and past chest-high stacks of books to his small desk on the far wall. You can't see the door from this space for all of the scientific papers that block your view. But if you ask for any paper in his collection, Dave will find it in the middle of any stack. I can barely manage to find my way around his stacks, but he knows where everything is inside them. This is no clutter of a disorganized person; this is the ideal arrangement for a mind capable of finding order in chaos.

Dave crunches databases to find signals in the history of life much as the Woods Hole group tried to do forty years before. He mostly looks at shelled creatures because they are abundant and readily preserved in the fossil record. Dave is inspired by the search for large-scale patterns. Every measurable feature is fodder for his analysis, such as how big the species were, and where and when they lived.

Removing the noise from the data is a tricky business. Let's compare hypothetical fossil species and ask a simple question: Which one was more abundant in the distant past? Start with the obvious. Count every fossil of those species ever collected in every museum, and draw the simple conclusion that the most abundant species in the past is the one that has the most fossils in the museum collections. But we'd quickly realize the big problem: some fossils are common because they preserve easily. Or they may be easier to find. Still others are common because collectors liked them disproportionately; maybe they were relevant to a particular project somebody was doing. If you were to look at our collection from the Arctic, it is heavily weighted toward teeth and the back ends of jaws. Does this mean that teeth and jaws were more common than the rest of the animal? Of course not. It only means that they preserve and are found more easily than other parts. Dave Jablonski and his colleagues spend a lot of their time trying to remove these biases and noise from the fossil record to find the real signal—the census of life on our planet over time.

Clams, oysters, mussels, and their relatives are not only features of the dinner table but also one of the most abundant components of the world's fossil record. Common in ancient lakes, streams, and oceans, bivalve species fill cabinet after cabinet in paleontological collections worldwide. Their wide distribution in the fossil record (they have been on the planet for over 500 million years) makes them an ideal laboratory to test theories on how species diversity changes through time.

To see things Dave's way, you need to think about the 3.5-billion-year history of life as one big survival game, where the creatures that live longest and produce the most fertile offspring win. Then think about the features that help species survive and reproduce. For animals, you'd likely make a list that includes traits like the abilities to run faster than predators, to jump high, to climb efficiently, and to have jaws specialized for particular foods. It might mean being big at some times and small at others. You could measure how well creatures do certain activities, such as feed, reproduce, and move about. You could use these measurements to make predictions of winners and losers: faster animals would out-survive slower ones, faster breeders would do the same to slower ones, and so on. And for large chunks of geological time, tens or hundreds of millions of years, these features would seem to relate to the success of different species. Then you'd look to see how these features help animals at the biggest catastrophes in the history of life. You'd guess these features would be keys to long-term success. And you'd be dead wrong.

KINGS OF THE HILL

What is the holy grail of paleontologists, the feature that predicts success during cataclysm? In the immense history of Earth, on many continents, over billions of years, through extinctions caused by asteroids, sea level change, and volcanic eruption, there seem to be rules about what happens to living things during cataclysms. One trait—among all those that life has ever had—seems to give us the ability to predict whether a species is likely to live or die at a catastrophe. The best survival tip for a species is to be widely spread around the globe. Species that have individuals spread about, preferably on different continents, fare better than those that are found in only one spot.

For millions of years, survival and reproduction depend on how well creatures feed, move about, reproduce, and so on. Then, every so often, a catastrophe happens, and those traits become virtually meaningless. What matters is the happenstance of where they live. Rare events wipe the slate clean and briefly change the rules of the game. The creatures that survive catastrophe aren't always "better" at any ecological trick. If the ultimate victory is surviving a catastrophe, then the winners are those that are globally distributed.

If creative destruction is good for economies, so too is it for the biosphere. Survivors of global calamities inherit a new Earth— a planet with fewer competitors. Imagine a game of king of the hill. A huge, mean playground bully sits at the top of the hill and, with the advantages of his elevation and size, simply owns it. Nothing you do can get you up there. What is the best gift you can be given in this game? Maybe some random event, perhaps his mom calling him home for dinner, leaving the hill open for you. With the bully gone, you can simply scamper up the hill and gain the advantage of elevation to use when others come up.

The king-of-the-hill idea may also hold for species. If a successful species occupies some niche, perhaps lives in a particular zone of the ocean, others cannot easily occupy that ecological

Each catastrophe leaves a line of survivors with a new Earth.

space. Now, if a cataclysm removes that ecological version of king of the hill, the survivors can occupy the prize position without so much as a fight.

From our perspective, as one species sitting on top of 3.5 billion years of life's history, we ask: What has this meant for us?

Most of our fossil hunts are spectacularly unsuccessful, and my work in the 1990s with Farish Jenkins in Africa was no exception. We spent a fruitless month looking for mammals in

200-million-year-old rocks in Namibia, and at the finale Farish wanted to boost morale by taking us up north on a safari. After a few days' drive, we found ourselves in Etosha National Park, a vast desert along the border with Angola. The desert plain is dotted with small water holes that are magnets for life. Every day we'd haul out of bed at sunrise, park our cars next to a quiet water hole, and sit for hours, simply watching the panoply of life come and go. First the birds arrive. Then come the zebras and buffalo. A pack of hyenas might wander about. Everybody scatters as a lion circles, then, when things seem safe, the whole crew settles back to a normal rhythm of feeding and drinking.

Here we were in a glorious world of large mammals and birds of all kinds, but my brain was still locked inside the patterns of rocks of 200 million years ago. At that time, reptiles of every imaginable description roamed Earth; mammals were tiny shrew-sized creatures, and birds were nonexistent. Daily life at the water hole contains the imprint of catastrophes millions of years ago. The water holes before that time were loaded with a different creature, a very successful one. Dinosaurs, large and small, plant eating and carnivorous, occupied these niches. Instead of elephants and large plant-eating mammals, in the Cretaceous there were herds of ceratopsians and hadrosaurs. In place of large lions, there were tyrannosaurs and other large dinosaurian and crocodilian carnivores. The dinosaurs and their cousins were the kings of the hill for eons until they got knocked out by catastrophe. Only then did the descendants of a little mouselike creature, with teeth as small as grains of sand—whom dinosaurs trampled under their feet—grow to become the new kings of the hill.

CHAPTER EIGHT

FEVERS AND CHILLS

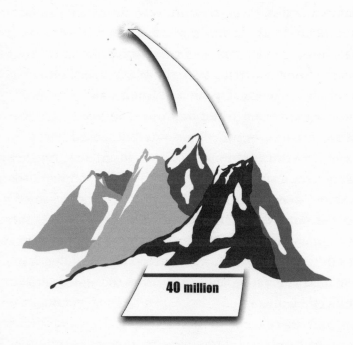

40 million

Arctic bush pilots are a special breed. Years of solo flying endow them with a crusty independence and deep familiarity with the landscape. After countless hours looking down on terrain, pilots' eyes can discern patterns hidden from the rest of us. During one flight in 2002, our pilot suddenly pushed the plane into a steep dive and veered into a tight bank, a maneuver that dropped us from ten thousand feet to about two hundred over the water in a small fjord. While I was seeing my life flash before my eyes, he saw a school of fish and, being a fisherman in his spare time, wanted to get a closer view. Even if my eyes weren't closed, there was no way I could have perceived swimming arctic char from such an altitude.

During one chopper run in 1985, a pilot named Paul Tudge was shuttling supplies between distant camps on Canada's Axel Heiberg Island and Eureka Sound, two of the North's most spectacular places. When the air is clear and the ground free from snow, the colors and images are so sharp that tiny details can be visible miles in the distance. In this part of the Arctic, barren mountain ranges border gentle valleys. The enduring action of ice, wind, and intense cold sculpts the bedrock into a range of obelisks, sheer walls, and potholes that almost seem unnatural. The sensation of otherworldliness is magnified by the lack of large plants: there are no standing trees, shrubs, or even grass in this area.

Scanning brown, gray, and red vistas below his chopper, Tudge noticed something odd on the bedrock floor. Wind had winnowed a depression, out of which poked objects that looked like tree stumps. Not believing that there are trees in the Arctic, let alone ones that grew out of rock, Tudge set the chopper down. Lying in wait for him were not only tree stumps but piles of branches, logs, and other tree parts jutting from the ground. He dutifully collected samples and shipped them to one of the Arctic's leading fossil plant experts, James Basinger of the University of Saskatchewan. Basinger dropped everything and mounted an expedition as soon as money and permits could allow; of course, in the Arctic this process can take a year or more.

Awaiting Basinger's spade was an entire buried forest mummified in eroding rock. The cold dry air left fine anatomical details of the leaves and wood intact, including their original cellular structure. The wood of these trees even burns. There is a big difference between these logs and a Duraflame; the Arctic ones come from a forest over 40 million years old.

The stumps that jut from this frozen landscape expose redwood trees that would have reached heights of 150 feet or more.

Tudge saw exposed stumps (left) that contained beautifully preserved wood over 40 million years old (right).

In the past, this place was no barren wasteland; it was alive with plants much like Northern California is today. Of course, nowadays the tallest tree up north is a little willow that rises mere inches off the ground. It is almost as difficult to see Arctic willows from six feet as Tudge's fossil forest from the air.

About twenty years before Paul Tudge's flight, the eminent paleontologist Edwin Colbert received a box in his office at the American Museum of Natural History in New York City. Sent by a famed geologist from the Ohio State University, it contained a sheet of official letterhead wrapped around an isolated bone the size of a human finger. The colleague had collected this fragment in the field and wanted Colbert's expert opinion.

From his many years on expedition to the American Southwest, Colbert was able to identify the bone in a split second: it had the distinctive texture and shape of a jaw from an ancient amphibian that lived over 200 million years ago. Looking somewhat like fat crocodiles, these creatures were widespread throughout the globe for a good chunk of geological time. But this ordinary-looking fragment was very special: it came from the Transantarctic Mountains, a range two hundred miles from the South Pole.

Colbert was a longtime fossil hunter, and the sirens went off in his head. Opportunity knocked; here was a continent completely unexplored for fossils. Colbert wasted no time assembling a dream team of experts from the United States and South Africa, who from years of working on the rocks of this age had the eyes to find new fossils. If fossil bones were present in Antarctica, this was the team to find them.

Almost from the moment their boots touched the Antarctic sandstones, Colbert and his team had a field day picking bones from the sides of barren hills, where fossils were virtually everywhere they looked. One creature had a body shaped like a medium-sized dog, only instead of a jaw like a carnivore it had a large birdlike beak. What stopped Colbert in his tracks was not this creature's bizarrely chimeric form but something far more mundane: paleontologists had known of this creature for decades. In the 1930s, South African geologists identified an entire layer that contains thousands of them extending across a wide swath of the Karoo Desert. This so-called *Lystrosaurus* Zone even reaches South America, India, and Australia. Now, with a *Lystrosaurus* level in Antarctica, Colbert and his colleagues uncovered yet another clue exposing the reality of continental drift. With the match of the rocks, coastlines, and fossils, a new view of Antarctica emerged: the continent in the past sat as a keystone at the center of a vast supercontinent that included Africa, Australia, and India. This clump of continents covered much of the southern part of the planet.

The pile of fossils that Colbert's group discovered revealed another fact of Antarctica's past. *Lystrosaurus*, like the amphibian that led him there in the first place, was a cold-blooded animal that could live only in warm tropical or subtropical climates; think big salamanders or lizards. Ditto the fossil plants. Colbert and his team labored near the center of a vast frozen continent, close to where Robert Falcon Scott and his party froze to death nearly sixty years before. But everything inside the rocks pointed to one conclusion: Antarctica was once a warm and wet world teeming with tropical life.

Expeditions that followed Colbert's only exposed more of Antarctica's disconnect between its frozen desolate present and its lush past. The world that Colbert uncovered was followed by another filled with dinosaurs and their kin. In rocks even more recent, 40 million years old, this tropical continent was home to modern rain forests, amphibians, reptiles, birds, and a whole menagerie of mammals. For most of its history, Antarctica was a paradise for life.

Then, starting 40 million years ago, the entire continent went into the freezer, and with it Antarctica witnessed the greatest and most complete extinction of any continent in the history of the planet. From a world rich in plants and animals, virtually every land-living creature simply disappeared.

There is symmetry to Tudge's flight near the North Pole and Colbert's exploration of the South: one unveiled temperate forests, the other tropical animals in regions that today house frozen deserts. The story of the poles is that of the entire planet. Our present—with its polar ice—is an aberration. For most of history our planet was warm, almost tropical. If the rocks of the world are a lens, they reveal that our modern, relatively cold landscape is not the normal state of affairs for the planet.

In this great cooling lies one of the major events that shaped our bodies, our world, and our ability to see all.

TAKING OUR TEMPERATURE

Carl Sagan once spoke of a paradox about our planet's climate. The sun is not a constant beacon of light; it started its stellar life as a relatively dim star over 4.6 billion years ago and has increased in brightness ever since, being about 30 percent brighter and warmer now than when it formed. With such a dramatic increase in heat over the years, Earth should have been a frozen waste in the past and now be a roiling cauldron of molten crust. Yet all our thermometers for the planet paint a different picture. Glaciers exist today during an era when the temperatures should be downright hellish. There are signs of liquid water inside 3-billion-year-old rocks—at a time when Earth should have been a ball of ice. Sure, we've had our moments of hot and cold, but if you think about Venus's surface temperatures of 900 degrees Fahrenheit and Mars's of -81 degrees, Earth has been a stable Eden relative to its celestial neighbors. Somewhere on the planet lies a thermostat that buffers it from dramatic extremes in temperature.

Inroads to the thermostat were discovered by a student who was as persistent as he was headstrong. He started his graduate career by loudly proclaiming to his thesis adviser that he had a brand-new theory of electrical conductivity. The response to this arrogant introduction was a simple "good-bye." Perseverance paid off, and, probably to the relief of his teachers, in 1881 Svante Arrhenius moved to Stockholm to work with a professor at the Academy of Sciences there. After that gig, Arrhenius went on to think of other scientific problems.

One scientific puzzle was right in front of Arrhenius's eyes. He saw the factories of the Industrial Revolution belching coal smoke—in his words, "evaporating our coal mines into the air." Arrhenius knew from previous work that carbon dioxide, a major constituent of the fumes, could capture heat. He made a few calculations that revealed how increased carbon dioxide in the air would trap heat on Earth and raise global temperatures. This idea was to lay fallow for a number of years, during which time Arrhenius won the Nobel Prize for work derived from the seemingly lackluster doctoral thesis that had so annoyed his professors.

The famous greenhouse effect is based on Arrhenius's work. The more carbon dioxide there is in the atmosphere, the more heat is trapped by the planet and the hotter things get. Of course the reverse is true. But there is a deeper meaning to carbon in the air, one that emerges only when you take the long view on timescales that extend millions of years in the past.

The television character Archie Bunker once famously said of beer, "You don't own it, you rent it." The same holds true for every atom inside us; we are the temporary holders of the materials that compose our bodies. Few of these constituents are more important to the balance of life and the planet than carbon. The connection among parts of Earth depends on how carbon moves through air, rock, water, and bodies. To see this chain of connections, we need to consider living things, rocks, and oceans not as entities in their own right but as stopping places for carbon as it marches along during our planet's evolution.

Viewed in this way, the amount of carbon in the air depends on a delicate balance of conditions. Carbon in the atmosphere mixes with water and rains down to the surface as a slightly acidic precipitation. We see the effects of this in our daily lives; in my university, built largely in the late nineteenth century, few gargoyles still have faces. Acid rain works on exposed rock everywhere—on mountainsides, rubble fields, and sea cliffs. Once the acid rain breaks down rocks, the water—now also enriched with carbon that was inside the rocks—eventually winds its way through streams and rivers into the oceans. At this point the carbon gets incorporated into the bodies and cells of the creatures that swim there: seashells, fish, and plankton. When sea creatures' remains, loaded with carbon, settle to the bottom of the ocean, they ultimately become part of the seafloor. And, as we've known since Marie Tharp, Bruce Heezen, and Harry Hess, the seafloor moves, only to be recycled deep inside Earth.

This chain of events removes carbon from the atmosphere, taking it from the air and moving it to the hot internal crust of Earth. Alone, these steps would pull all the carbon out of the air, leaving Earth freezing with no atmospheric insulation. The good news is that there is a recycling mechanism for carbon. Carbon in the interior of Earth gets injected back into the atmosphere by

The cycle of carbon. Over long time periods—millions of years—carbon enters the atmosphere as it ejects from volcanoes, only to return to rock at the bottom of the ocean so that the cycle can begin anew. Acid rain is the link: it takes carbon from the air, allows it to drain into the ocean, and get incorporated into rock.

volcanoes that eject gases. That is the long-term source of much of the carbon we breathe: while acid rain and the weathering of rocks remove carbon from the air, volcanoes emitting gases return it. Volcanoes typically release huge amounts of water vapor, carbon dioxide, and other gases; by some estimates they send over 120 million tons of carbon dioxide into the air each year.

Like a sequence of events in which each step makes sense but the end points are counterintuitive, the conclusion to draw from carbon's movement is that rock erosion and weathering is linked to climate. Rock erosion by acid rain is like a giant sponge that pulls carbon dioxide from the atmosphere. Lowering the amount of carbon dioxide in the air will drop the planet's temperature. On the other hand, planetary events that increase the amount of carbon in the air—enhancing volcanic activity or slowing removal of carbon from the air—will, of course, serve to raise temperatures. All else being equal, increasing erosion of rocks leads to lower temperatures, decreasing erosion to higher ones.

The movement of carbon links rocks to climate and ultimately answers Sagan's paradox about the sun. The planet's temperatures are kept within a narrow range by the movement of carbon molecules through air, rain, rock, and volcano. Hot weather leads to more rock erosion, which leads to more carbon being pulled out of the air and thus colder weather. Then, just as things get colder, the cycle moves the planet's temperatures in the opposite direction: colder weather leads to less erosion, increasing amounts of carbon in the air, and hotter temperatures. Liquid water is possible on our planet only because of this balance; neither we nor the landscapes we depend on could exist without it. But liquid water is like the miner's canary. Too much of it, or too little, reveals a long-term shift in workings of the planet, changes that amount to planetary fevers and chills.

What happened when the poles started to freeze about 40 million years ago? The shift from hot to cold occurred at the same time that the levels of carbon in the atmosphere dropped precipitously. But this begs the question: What changed the levels of carbon in the air?

Maureen "Mo" Raymo went to school to study climate and the kinds of geological changes that could have an impact on it. And, like Arrhenius, she produced a thesis that elicited memorable comments from advisers. One went so far as to comment that her Ph.D. dissertation was "a total crock."

Her path to that fine moment began like any other graduate student's: she took a string of classes representing the core knowledge of her field. In geology seminars of the 1980s, much of the buzz was about global carbon and Earth's thermostat. A classic paper, read by every student at the time, written by Robert Berner, Antonio Lasaga, and Robert Garrels, described this link in chemical detail. The paper became affectionately known as BLaG after the initials in the last names of each author. Everybody read BLaG and everybody was tested on BLaG, even though virtually everybody, including the BLaG authors themselves, realized key details of their brilliant model had yet to be filled out.

Raymo took the standard class for graduate students where the details of BLaG were presented. She also took classes on modern rivers, mountain formation, and tectonics. But unlike the rest of us who sat through this kind of curriculum, she began to connect the dots.

Everybody knew that the climate cooled drastically starting 40 million years ago, but there was no known geological mechanism that could possibly have done this. What could drop the temperatures? Only a major planetary change could possibly have removed enough carbon to allow such cooling.

Then Raymo looked at a globe and remembered her plate tectonics. The period of drastic cooling commenced at a pivotal time in the history of the planet. This was when the continental plate of India, which had been traveling north for hundreds of millions of years, began to slam into Asia. The result of this collision is like sliding two stacks of paper along a tabletop until they scrunch together: they crinkle and rise. A similar kind of mashup of the continents led to the rise of the Tibetan Plateau and the Himalaya mountains.

Raymo's lead adviser (not the one who called her thesis a crock) was thinking about how a new mountain range would affect global wind currents or serve to make a shadow that could foster storms. Raymo's insight came from thinking about how a massive mountain range and plateau could affect Earth's thermostat.

The Tibetan Plateau is a vast barren face of virtually naked rock. It contains over 82 percent of the rock surface area of the planet and reaches over twelve thousand feet high. With the rise of such a plateau came ever growing amounts of rock erosion on its surface. When we look at the Himalayas, most of us see a dramatic series of mountains, but Raymo saw a giant vacuum that removes carbon dioxide from the atmosphere—and rivers that flush the carbon into the sea. With decreasing carbon in the atmosphere came a cooler planet. The rise of the Tibetan Plateau led to the shift from a warm Earth to a cold one; it did so by pulling carbon from the air via erosion of rock.

Raymo's theory makes sense of an enormous amount of data, but gaining support for an idea like this is more like winning a criminal case on circumstantial evidence than it is a mathematical proof; only the agreement of a heap of independent lines of evidence can nail the case. Raymo made a very specific prediction: the test lies in using tools that can correlate measurements of the rates of uplift of the plateau—and the levels of weathering of rock—with the amount of carbon in the air. There are altimeters in ancient rocks—altitude-sensitive plants. Carbon in the air dropped at the time of uplift, but we still do not have the precision to tie the different variables together in the fine detail needed for a test. Whether erosion of the plateau alone is sufficient for the climate change we see or if this acted in concert with other mechanisms remains to be seen.

By 40 million years ago, the map of the world was on the move, and with it the environment that supports life. India slamming into Asia may have heralded an era of dropping levels of carbon dioxide and global cooling, but details of the timing of the freezing of Antarctica suggest other contributing factors. Forty million years ago, it went from being a rain forest to having a climate much like southern Patagonia today. Then, by about 30 million years ago, the fauna and flora started to diminish, until 20 million years ago, when the first permanent sea ice appeared. The vegetation was largely that of stunted tundra by this time. Ten million years ago desolation was complete.

Look at a map, and you'll notice that the Northern Hemisphere is mostly brown land and the Southern Hemisphere is mostly blue ocean: the northern half of Earth is composed of large and mostly connected continents, and the southern one, vast oceans. Inside this simple observation lie clues to the freezing of the planet, the vanquishing of life on Antarctica, and the environmental changes behind much of human history.

By the early 1970s, as the reality of plate tectonics was being widely acknowledged, one huge patch of ocean floor remained virtually unexplored: the waters of the southern oceans made famous by the great explorers of Antarctica such as Robert Falcon Scott, Ernest Shackleton, and Roald Amundsen. Rough seas separate icebergs and barren rocky islands, so much so that the latitudes from 40 degrees to 70 degrees in which this southern ocean sits are given nicknames: the "roaring 40s, furious 50s, and shrieking 60s." This was among the last regions of the ocean floor to be sampled for a good reason. The ocean currents and winds make for a forbidding sea.

Coming off their successes mapping the floors of the Atlantic and the Pacific, the ocean-mapping teams that supported the work of people like Heezen and Tharp turned to this southern patch. From 1972 to 1976, expeditions were sent to core about twenty-six sites to look at the sediments of the ocean bottom. At each stop in the ocean, at a site determined by studying maps at home, a winch returned a plug of seafloor captured from a core at the bottom. Each rock was studied chemically to reveal its age and origins, and the structure of the ocean bottom was mapped just as Marie Tharp and Bruce Heezen had done a decade before in the Atlantic.

The cores from the sea bottom changed the way we look at the southern world. Surrounding Antarctica like a ring is a huge rift valley with a molten core. This, like the rift at the bottom of the Atlantic, is a place where new seafloor is being made, where the plate is actually spreading. The jigsaw puzzle of the South, contained in the shape of the continents and Colbert's *Lystro-*

saurus discovery, became clear. At one time in the past, the entire southern part of the globe was indeed one giant super landmass composed of what are today all the southern continents: Antarctica, Australia, South America, and Africa. The distinctive blue oceans that define our South today weren't there.

Then, with the birth of this volcanic ring surrounding Antarctica, the continents separated and moved away from their southern neighbor. Three things happened at once: Africa, Australia, and South America moved north, Antarctica became isolated at the South Pole, and vast seas opened up separating all the southern continents. None of these changes boded well for life near the South Pole.

Just as isolation is bad for people, so too is it for polar continents. The ocean current that has so vexed mariners for years runs as a ring around Antarctica from east to west. It was born as space was created for it by continents cleaving from Antarctica. Oceans are wonderful ways to transport heat. As an example, Britain lies at the same latitude as northern Labrador. One place is relatively mild, the other quite fiercely cold. The reason? Warm currents coming up from the equator keep Britain's climate mild, whereas the western Atlantic has no such current. Before Antarctica became isolated, ocean currents running from the equator brought heat to the continent. When Antarctica separated from the other southern continents, this conveyor of heat stopped, only to be replaced by the ring current. This change spelled cold for Antarctica: whatever heat existed at the South Pole just escaped into the air, never to be replenished by warm waters. Life on Antarctica literally froze to death or skedaddled to greener pastures elsewhere.

The emerging map of the world changed climate and life. Moving continents and expanding seafloor brought new patterns of ocean circulation, erosion, and levels of carbon dioxide in the atmosphere, thereby dooming an entire continent. The consequences extend as far as the eye can see.

SEEING IT ALL

Humans are visual animals built to detect patterns in a messy world. Bush pilots' eyes, like those of Paul Tudge, are trained to spot objects on a flight. Children can find hidden objects in puzzles or pictures, fly fishermen learn the water by seeing shadows below the ripples in streams, and radiologists save lives by deciphering shadows on images: our species has survived by finding patterns hidden in the apparent chaos around us. This ability lies in the interplay between our eyes and our brains: together they help us learn to see, survive, and thrive.

We live in a world so awash in vivid hues that it is easy to forget we perceive only a tiny fraction of the colors in front of our eyes. Light arrives to us in a wide spectrum of wavelengths, from ultraviolet to infrared. Gadgets such as night-vision goggles provide only a glimmer of these hidden frequencies. Other creatures can see a broader range of colors naturally. Birds perceive many more shades of blue, as do some species of fish. Each species—whether eagle, trout, or human—is tuned to experience and perceive its world in a particular way. And our perception has its roots in the forces that froze the poles of Earth.

Human eyes, like those of other mammals, have a postage-stamp-sized retina in the back that receives light from the lens. Plastered on the retina are about 5 million specialized cells that are like little receivers to detect red, yellow, and blue—the three primary colors of light. This ability is conferred on each cell by a specialized protein inside that undergoes a distinctive change in shape when the right color hits it. The cells in the retina can discriminate about a hundred different shades of light. When these signals hit the brain, they are combined, allowing us to perceive a palette of about 2.3 million different colors.

Our closest primate cousins of the Old World—monkeys, gorillas, chimps, and orangutans—can see the same palette of colors as we do. We share a very similar makeup, which extends to the proteins inside the retina we use to perceive color. More distantly related primates, such as those that live in South America, do not see in color exclusively: in some species the males are color-blind. Ever since the nineteenth century, primatologists have known about a big split in our primate family tree: all Old World monkeys have full color vision, whereas this trait is lacking in their New World cousins. Is there also a difference in lifestyle that explains the ability to see in vivid color?

The first hint came from a surprising finding. Howler monkeys, as the name implies, have a distinctive cry. They were described by the great explorer Alexander von Humboldt in the nineteenth century as having "eyes, voice, and gait indicative of melancholy." Scientists studying their behaviors and visual structures in the 1990s discovered that unlike South American monkeys, all howlers are able to see in the same spectrum of color as we do. There is a huge difference in the diets of howlers and their South American cousins. All other monkeys eat mostly fruit, whereas howlers exist on leaves.

This observation motivated a young graduate student, Nathaniel Dominy, a former football player at the Johns Hopkins University, to think in a new way about how color vision arose. Perhaps the lessons of the howlers is general, he thought, and there is a major difference in diet that explains why our branch of the primate family tree sees in color.

Kibale National Park in western Uganda sits among a rich forest landscape of evergreen and mixed deciduous trees. Leopards, hornbills, and distinctive forest elephants—unusually hairy and small—dwell there. So too does an astounding diversity of primates. A whopping thirteen species—including chimpanzees—make the park their home.

Kibale is also home to a fourteenth species of primate—humans—many of whom live in the Makerere University Biological Field Station to study their primate cousins. In 1999, Dominy traveled there with the simple goal of watching the monkeys eat.

Dominy and his research adviser, Peter Lucas, had a plan: they were going to look at each type of primate in the reserve and quantify exactly what it ate and when. If there was a pattern to the diets, they were going to find it. The crew wasn't just armed with notepads; they carried a backpack laboratory that was described in a later scientific paper with a title that says it all: "Field Kit to Characterize Physical, Chemical, and Spatial Aspects of Potential Primate Foods." Inside the backpack was a materials testing device designed to measure toughness of foods; a spectrometer to quantify color and basic nutritional properties of foods; and a number of other gizmos to record the shape and weight of whatever the monkeys gobbled up.

Dominy, Lucas, and their team spent ten months watching primates. When they weren't interrupted by the threat of bandits or terrorists (at one point they were forced to retreat to the American embassy in Uganda), they worked around the clock, eventually logging 1,170 hours of observations. They would watch the animal as it consumed its meal, then hit the leftovers with their backpack lab. In the end, they found that the monkeys consumed about 118 different kinds of plants.

Back home, as they crunched the data, a pattern emerged. Species with color vision preferentially selected leaves that varied on the red-green color scale. That is, they were differentiating foods that animals lacking color vision could never even perceive. And what of the foods that they selected using their color vision? These morsels uniformly had the highest amount of protein for the least amount of toughness. The primates' mothers must have been pleased: they ate things that were both good for them and easy to digest. And the biggest cue, red color, was something that only species with full color vision could detect.

To Dominy and his colleagues, a hypothesis emerged: color

vision enabled creatures to discriminate among different kinds of leaves and locate the most nutritious ones. This advantage gained new prominence when climates changed and plants responded.

More clues to color vision are nestled inside DNA. Mammals that lack color vision have only two proteins to perceive color; we and the Old World apes that perceive colors have three. In 1999, as DNA technology became cheaper and more powerful, the actual composition of these proteins could be compared, giving a detailed look at their chemical structure. Hidden inside the sequences was a major clue to the origin of color vision. The three proteins that allow us to see colors are duplicates of the two seen in other mammals. By comparing the sequences in the new copies with the old ones, we can get an estimate of when the duplication happened. All creatures with the three genes trace their lineage back to about 40 to 30 million years ago, the likely time when color vision arose in our closest ape ancestors.

What happened to the planet at the time of this genetic change? Earth got cooler. The forests of Antarctica and the North retreated, to be replaced by ice. Grasses spread to new places around the world. The fruit-bearing palm and fig trees, so common in Wyoming and throughout the warm world, declined, yielding forests mostly of leaves—some tough, some soft, some nutritious, some inedible. The skills that are now so useful to the primates with color vision in Kibale, Uganda, were a key to their success during the period of global cooling. The cold brought a new flora, one that put a new kind of color vision at a premium.

Paul Tudge discerned tiny stumps in a vast landscape, paleontologists find tiny fossils inside a field of rocks, and our primate ancestors survived climate change by discriminating nutritious foods in a dense collage of leaves in forests. Every time you admire a richly colorful view, you can thank India for slamming into Asia, continents for retreating from Antarctica, and the poles for becoming frozen wastes. Buried within it all lies the way carbon atoms move through our world.

CHAPTER NINE

COLD FACTS

10,000

As we lumbered along at a ground speed of thirty miles per hour, I felt as if our plane would drop from the sky. With this strong headwind, the five-hundred-mile trip from Iceland's capital, Reykjavík, to our remote landing strip on east Greenland could take nearly half the day. The craft we were flying, a DeHavilland Twin Otter, is the workhorse of the Arctic. With a stall speed of fifty-five miles per hour and fitted with huge balloon tires or skis, it can land on tiny patches of rocky tundra or ice nestled in remote Arctic valleys. Hunched in a compartment big enough for only four crew, pilots, and gear, I could only imagine how early Arctic explorers—those who set out in the nineteenth century with wool coats, leather shoes, and salt pork dinners—felt on their first sight of the North. With the slow ride and a window seat, I was able to linger on the view.

During the trip north, the vista transforms as plants recede and the extent of ice expands. The sea ice appears in patches at first, then forms a solid sheet over the ocean. Seen from ten thousand feet, the ice grades from a clean, almost pure white to shades of blue, green, and teal. With shapes like no other place on Earth, it fractures in cubes in some spots and in long sticks and crystalline diamonds in others.

The slow, low-altitude approach to Greenland is defined by an ominous wall of fog that lingers in the windshield for hours. As one gets closer, the fog is revealed to be a massive sheet of ice that extends as far as the eye can see. The center of the island is filled by one of the largest glaciers on the planet. The sheet extends six thousand feet high and six miles deep, over an area the size of Texas. Exposures of bedrock of the island are restricted to cliffs that line the coast; the rest of Greenland's rock lies buried deep under ice. The ice cap is a lifeless glacial desert touched by humans only rarely.

In one bout of activity, this desert of ice sprang to life in the 1950s, when Greenland took strategic importance during the Cold War. On the northwest corner of the ice cap a secret project run by the U.S. Army was launched with a name right out of *Dr. Strangelove:* Project Iceworm.

The plan, concocted somewhere in the seventeen miles of Pentagon hallways, was to carve silos for six hundred nuclear warheads in the ice in northern Greenland. Connecting these silos were to be tunnels that contained an entire underground city given the futuristic name Camp Century.

The project was started secretly in 1959, when twenty-one tunnels were dug with heavy equipment flown from bases far to the south. In its heyday, this city under the ice housed over two

In its heyday, "Main Street" of Camp Century was eleven
hundred feet long. The area was crushed by ice by 1969.

hundred people and contained a shop, hospital, theater, even a
church. Power was supplied by the world's first portable nuclear
reactor, the plucky Alco PM-2A. Heat from the reactor melted
ice to provide the subglacial city with water. Self-sufficient and
mostly belowground, the whole operation was something like a
human ant farm.

Being close to perceived threats in the U.S.S.R., Camp Century had the makings of a perfect military base. The plan worked
well, except for one problem: ice moves. By 1966, it had become
clear that the ice was shifting so extensively that warping tunnels
would destroy expensive equipment. Pictures of Camp Century

taken today reveal twisted machinery and abandoned huts, all artifacts of schemes and fears inside an ancient block of ice.

Work at Camp Century did have value, though not of the type Pentagon planners could have ever foreseen.

A SUMMER VACATION THAT CHANGED THE WORLD

Louis Agassiz was born in 1807 with charm, intelligence, and an unrelenting passion to study nature. Even as a child, he fed his insatiable curiosity by making his own collections of animals and plants, often drawing each of their organs in exquisite detail. He believed in learning by seeing, a dictum that was to become his catchphrase throughout his career. Sensing his proclivities at an early age, his parents set him up to apprentice with an uncle who had established a successful business. They wanted Louis to develop into a successful "man of affairs," not a collector of bugs and rocks. But they underestimated the influence of his charm. Young Agassiz would have none of his parents' designs: he enlisted one of his teachers to lobby his parents for him to stay in school and become, as he said later, "a man of letters."

While Louis was in his late teens, he and his brother were studying in Zurich and found themselves without a ride home, a distance of over a hundred miles. They started walking until a stranger, a well-to-do Swiss, offered them a lift. So impressed was he with Agassiz's acumen that this gentleman later wrote to his parents offering to pay for his full education. Thus began a steep career trajectory that ultimately propelled him to the United States, where he took part in the founding of two major scientific centers: the Museum of Comparative Zoology at Harvard and the National Academy of Sciences.

As a young family man in 1837, Agassiz took his brood on a summer vacation to the picturesque town of Bex. Lying along the Rhône River, Bex is bordered on the east and west by the

Alps. Today it is home to the only working salt mine in all of Switzerland. A
narrow-gauge train takes visitors hundreds of feet beneath the earth. This vast
hole was originally dug in the 1820s to
quarry salt that in those days was literally
worth its weight in gold. At the time of
Agassiz's visit the mine was new, and its
director took great pleasure in showing
summer visitors the local geology, which
in this part of the Alps is hard to miss
and very easy to appreciate.

Louis Agassiz.

Some time before Agassiz's visit, the director and a friend
had discovered a number of puzzles in the local rocks. With the
arrival of Agassiz, the two were excited to quiz a visiting lumi-
nary on the meaning of these geological oddities.

Giant boulders dotted the landscape, some the size of a
caravan. That is not unusual; large boulders can be a common
occurrence. But these were completely out of place, because the
rock that composed the boulders was different from the local bedrock. In fact, the closest match to the boulders was in cliffs hundreds of miles away. Something had transported them, but what?

Closer inspection of the boulders revealed other clues. Scrape
marks, almost as if made by a pickax, etched their surfaces. And
the marks didn't run willy-nilly; they extended in parallel lines.

More mysteries came from a bird's-eye view of the valleys
from the scenic overlooks that lie along the roadsides of this
part of the Alps. Each of the mountainous valleys was bordered
by ridges of gravel that looked scrunched, almost as if they had
been moved by a plow or a steam engine. Since the ridges were
perched on hillsides in rural valleys, these causes were obviously
ruled out.

Boulders and gravel mounds told the same story: something
was moving rocks around. But what?

Flowing water could be ruled out. Floods large enough to move the giant boulders would have left very obvious markings across the landscape. Of course, human activity could be ruled out as well. That left the one obvious cause—ice.

At the time of Agassiz's visit, the ice was nestled in glaciers high up on the mountains. But what if that was only its most recent position? What if at some point in the past the ice covered the valleys below? If the levels of ice waxed and waned in and out of the valleys, then the boulders would move, and the rubble would be plowed about to make mounds and carve scrapes.

After this grand show-and-tell, Agassiz's friends tried the ice idea out on him. To Agassiz—whose life's modus operandi was to learn by observing—the visit sparked an epiphany. It was a set of observations that changed his world. At every scale, Switzerland's rocks made sense when considered in the light of moving ice: scrapes on rocks told the same story as the mounds of gravel and the shapes of the valleys themselves. Agassiz's heart raced at the thought of something even more general. His travels revealed these features weren't limited to the Alps; they were common all over Europe, even south to the Mediterranean. Moving ice wasn't confined to picturesque Swiss cantons; it must have covered virtually all of Europe.

Unbeknownst to his friends in Bex, Agassiz set off to test his grand idea. In 1840, he published a book, dedicated to his friends from that fateful summer vacation, called *Studies on Glaciers*. In it he proposed the radical notion that ice at one point in time extended from the North Pole all the way to the Mediterranean and then retreated, only to extend again. A friend came up with a catchy name for these cold intervals: "ice ages."

Agassiz, with his personal charm, set off to convince the great eminences of the time of his notion. He took visitors out in the field as his friends from the summer vacation had done for him, encouraging them to see a past rich in ice. It took many trips, and even more arguments, but Agassiz succeeded. The ice age theory became widely accepted.

The beauty of this theory was that, like most great scientific ideas, it made specific predictions. Agassiz's notions could be tested simply by looking at the rocks in the world. Exotic boulders, mounds, and linear gashes on rocks should be widespread. If it is one thing to find a widespread pattern, then it is the clincher to find the cause.

But a problem for enthusiasts was that Agassiz's ice ages lacked any plausible mechanism. In fact, the idea even flew in the face of existing dogma that Earth has been cooling over time. If Earth was cooling, then glaciers should not have retreated to where they are today; they should have expanded. Moreover, Agassiz's layers of gravel and boulders were showing not a single shift but a rise and fall of Earth's temperatures over time. What caused the waxing and waning of the ice?

DANCING WITH THE STARS

Born and raised on a farm in Scotland, James Croll (1821–1890) lacked any formal education. Like Agassiz, he lived for the life of the mind: great ideas, puzzles, and intellectual problems. To support himself, he tried selling insurance, but with a natural aversion to people he couldn't stomach the job. Leaving that, he set up a tea shop. While he still couldn't manage to avoid people altogether, the shop did offer one salient advantage over the other gig: it left him plenty of time to study. And studying was the one thing he absolutely loved to do.

Croll's physiognomy, revealed by the best-known picture of him, shows the thousand-mile stare of one whose mind is transported to a faraway place or working on a deep mathematical problem. His mouth, set firm with a Scottish obduracy, also

James Croll (clearly not thinking about tea).

reveals a decided lack of humor; one can't imagine many jokes emerged from those lips. By all accounts, Croll had an exceptional focus that, coupled with a passion for learning, would allow him to spend an entire year reading a single book, often lingering on one page for a day or more to digest each idea. His driving passion was to get to the bottom of intellectual problems. Not satisfied with seeing only patterns, he wanted to figure out how the world actually worked.

Agassiz's ice ages provided a puzzle ripe for the solving. Croll's approach was decidedly different from that of Agassiz before him. Thinking of fundamentals, Croll asked, "What was the cause?" He set off with a pad and pen to solve the problem. His search for a cause demanded thinking about the factors that changed the amount of heat on Earth. The source for much of that heat is the sun. Is there some regular variation in heat from the sun that could trigger ice ages?

Soon after launching into this research, Croll read a paper by a brilliant French scientist that set his mind in motion. The idea was that regular variation in Earth's orbit could change the amount of heat that hits Earth's surface. Earth spins around the sun, and its tilt brings the seasons. The orbit depends on the proximity of other big celestial bodies nearby: Mars, Jupiter, Venus, and Saturn are all rotating in space as well. As they approach Earth on regular cycles, their large masses warp the orbit and tilt of our planet. In times on the order of thousands of years, Earth's orbit will wobble and change, thereby influencing the amount of sunlight that warms the planet. Croll reasoned that ice ages happen during regular intervals when the orbit causes the planet to receive less heat from the sun.

Here was a cause that made a
specific prediction: the ice ages
should happen at regular intervals defined by the orbit of the
planet. Unfortunately for Croll,
his theory became just a passing
fad. Because he lacked any firm
way of matching the timing of the
ice ages to orbits, Croll's theory
remained just a good idea.

Milutin Milankovitch.

A few decades after Croll's
death, a young Serbian concrete
engineer got the notion that he could use the mathematical tal-
ents that were so helpful in designing buildings to uncover how
the universe worked. His thinking was revealed in a toast he
gave a poet friend after the two shared a bottle of wine in a
Belgrade café. The poet had hoisted his glass to proclaim, "I
want to describe our society, our country, and our soul." The
concrete engineer countered with the salute, "I want to do more
than you. I want to grasp the entire universe and spread light
into its farthest corners."

Soon after the boast, the engineer, Milutin Milankovitch,
switched jobs. Leaving his building firm, he took a professorship at the University of Belgrade. Not easily intimidated, he
proceeded to announce that he was out to solve the problems of
the planet by pure mathematics. Global climates were his first
problem. But not just Earth's. He wanted to devise a mathematical theory for climate all over the face of Earth and for every
other planet in the solar system as well.

This ambition puzzled a few of his colleagues. Why would
you need to calculate global temperatures if we can simply set
up weather stations to measure them? Milankovitch's answer
revealed his thinking. If, armed with only pencil and paper, he
could predict temperatures mathematically, then we would truly

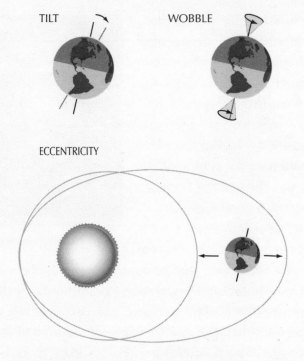

Milankovitch cycles consist of changes in the tilt of Earth, its
wobble, and the shape of the orbit around the sun.

understand their causes. Off he went, looking at the planetary
rhythms that so captivated Croll.

Croll's ideas were a natural starting point, but Milankovitch
brought a huge new twist to the problem. Using orbital calculations similar to Croll's, Milankovitch explored how sunlight
could change the heat of the planet. To elucidate this relationship, he modeled the different ways that heat gained by the ocean
can be transferred to the atmosphere and back. A brilliant math-
ematician, he was able to calculate the magnitude of temperature
changes during the seasons, resulting in a remarkably specific set
of predictions.

Earth's orbit changes in three major ways. Over 100,000 years
Earth's orbit goes from the shape of an oval to a more circular
pattern. During 41,000 years Earth rocks back and forth about

2 degrees. And in the course of 19,000 years Earth's tilt wobbles like a top.

Milankovitch realized that these are not huge changes, and in fact they would not alter the amount of heat received by Earth much. What they could do, as his equations showed beautifully, is change the duration and intensity of the seasons. And the reason is straightforward: if the seasons depended on the degree Earth is tilted and the manner in which the planet rotates around the sun, then changes to the shape and orientation of the planet and its orbit will affect the heat of summer, the cold of winter, and everything in between.

The rocks reveal the occurrence of ice ages. Mathematical calculations show that Earth's climate can change in a cyclic manner, matching the orbital changes of our planet. But do the cycles of ice ages and those of Earth's orbits march together? Answers would have to wait for new scientific quests—namely, the effort to make the atom bomb.

CHILLING EVIDENCE

The Manhattan Project was a short-term war effort that pulled together a unique cadre of scientists to focus on a single goal. With the war's end the U.S. government found itself with a problem, but one of those problems that is good to have. It had teams of scientific geniuses housed in different places, from New Mexico to New York, with no long-term infrastructure to continue their work. To make matters more challenging, no longer was there a single goal to their work, like developing a bomb; there were now many. Not wanting to lose the talent, or the momentum generated from fundamental breakthroughs in physics, the government supported a number of labs around the country, including one at the University of Chicago. Chicago was home to the group, led by Enrico Fermi, that launched the first controlled nuclear reaction (today the spot is marked by a Henry Moore sculpture across the street from the gym). After the war, the government helped the university establish a number of institutes exploring the big questions of physics and chemistry. One of those big problems was the history of our planet.

Two people who benefited from this transition from war to peace science at Chicago, Willard Libby and Harold Urey, shared a passion and a belief. The passion was for expanding knowledge. The belief was that trapped in the dynamics of single atoms—in their electrons, protons, and neutrons—were clues to the origin and history of the planet and perhaps even the entire solar system.

Driving this exploration of the atom was the development of new devices that could measure particles in parts per billion. With this resolution, new kinds of answers to old questions were now possible.

Libby set up two junior scientists in his lab with five thousand dollars to carry out a research program on carbon. Like most atoms, carbon exists in a number of different forms in the natural world. All carbon atoms have the same number of protons inside their nuclei; the different versions are distinguished from each other by the number of neutrons inside. Libby's insight was that all living things will have the same amount of carbon 14 in their bodies as the atmosphere in which they live. Living creatures breathe, eat, and drink carbon atoms in their daily lives and thus share the same balance of carbon with the atmosphere. Once organisms die, this balance with the atmosphere is disrupted, and no new carbon enters as food or nutrients. Whatever carbon atoms remain in the body begin to decay into other forms. As we've seen with other atoms, this reaction happens at a constant rate set by the laws of physics and chemistry. Knowing this, Libby ventured that if you can measure the amount of carbon 14 in a sample of old bones, you can, with some assumptions, calculate how long ago the animal died. This was a huge advance: it was like finding clocks inside ancient bones, teeth, shells, and wood.

To Harold Urey, who worked in a lab just steps away, atoms were imagined to be clues to the history of the planet, solar system, and universe. One of his main objects of fascination was an atom familiar to us all—oxygen. An abundant player in our air, water, and skeletons, oxygen has some distinctive properties that make this infinitesimal atom a window into our past and a much larger world.

Urey knew that oxygen, like carbon, exists as heavy atoms with extra neutrons and light atoms with fewer. On purely theoretical grounds, he guessed that the balance of these forms in any substance depends on temperature. The timing of his guess could not have been better, because accurate machinery could test his ideas.

And it worked: the ratio of heavy and light oxygen atoms in a material was dependent on temperature. To Urey and his team, this success meant that if you could measure the infinitesimal amounts of the different forms of oxygen in any substance— water or bone, for example—you might be able to guess the temperature of the environment in which it formed. The trick was to find the right kind of record that could reveal the details of Earth's climate with precision. Only then could the tool kit derived from the work of Libby, Urey, and their colleagues pull together cause and effect.

Seashells are durable and hard because they contain a crystal, calcium carbonate. This molecule, so vital to their hardness, also fortunately contains oxygen. Urey and others saw that as seashells develop during the life of the animal, the molecules that make the shell are ultimately derived from the water in which they lived. The relative amounts of the different forms of oxygen in the shell could, then, reflect the temperature of the waters that the creatures grew in. And since shells preserve well, they could contain an excellent record of ancient events.

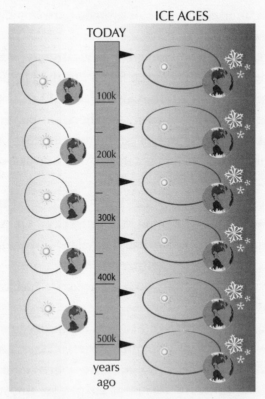

The 100,000-year cycle relates to changes in the shape of Earth's orbit: ice ages tend to occur more in eccentric periods.

With oxygen atoms as the thermometer, carbon atoms as the timekeeper, and the regularity of the layers as a guide, the teams set off to see how climate changed over the ice ages. One group looked at the most continuous record of seashells they could find, to map the temperature changes over time. The bottom of the sea is ideal: it contains layer after layer of sediment that drifts down the water column. By looking at the oxygen composition of the seashells inside these layers, the researchers could get an approximation of how climate changed over time. The team found that the planet's temperatures waxed and waned with peaks of high temperature and valleys of low temperature. What's more, the temperature seemed not to change randomly over time: if you squinted really hard at the graphs they made, you could see that the peaks and valleys seemed to rise and fall every 100,000 years. This was not some random number but one of those proposed by Milutin Milankovitch years before. One-hundred-thousand-year pulses started cropping up in other people's data as well. Maybe astronomical events were influencing things after all?

The problem was that the data were messy; the plots of temperature versus time have lots of wiggles, not just the 100,000-year one. Then three scientists, one British and two American, took a new look and applied a method developed by one of Napoléon's regional governors after his conquest of Egypt. The bureaucrat, bored on the job, set off to understand heat and its transfer among different materials. It wasn't heat that was to help geologists over a century later; it was a new mathematical approach he devised. If you have a graph with lots of different wiggles in it, perhaps that mess is made by several different rhythms superimposed on one another. The mathematical technique, known as Fourier transform analysis, is a way of revealing how a complex pattern can be made by a number of regular and more simple ones.

With that simple analytic tool, the data revealed not chaos but a deeply buried signal. The pattern emerges from a number of rhythms superimposed on one another: 100,000-year cycles onto cycles of 40,000 and 19,000 years. Milankovitch and Croll were right: ice ages are correlated in a broad way to the changing orbit, tilt, and gyration of Earth.

Graphs of climate, with peaks and valleys reflecting the rise and fall of temperature over the millions of years of geological time, look something like an EKG of a human heart. The heartbeat of our planet has drummed on for countless eons, beating to rhythms in Earth's orbit and the workings of air and water. Before the global cooling 45 million years ago that so fascinated scientists such as Maureen Raymo, these orbital changes did not often lead to ice ages. With a newly cool Earth, orbital wiggles became written in the waxing and waning of sheets of polar ice. And it is the ice itself that reveals the biggest surprises.

In 1964, during the heyday of Camp Century, a Danish geologist, Willi Dansgaard, visited the major air base in the region, Thule Air Base—the supply station for the camp—to look at local snow. Dansgaard spent some time in Chicago, even working in Urey's lab. Students then remember his fondness for the cold, leaving windows open during the long Chicago winters.

While on base, he heard buzz of the military project going on a hundred miles to the east. Asking permission to visit Camp Century, he was rejected on the grounds that it was a top secret operation. With some luck, in the form of a visionary senior administrator in the U.S. Army's Cold Regions Research and Engineering Laboratory, he was given access to the pristine cores of ice that the air force dug up to make the city under the glacier. Perhaps within these chunks of ice were keys to understanding the planet's climate?

Dansgaard had yearned to see a huge uninterrupted column of ice for much of his professional life, and now the most complete ice cores yet known were within his grasp. Two features of ice cores are immediately apparent. They are colorful, varying from iridescent green to blue. And they are layered, with thick layers, thin ones, and everything in between. Almost anything in the atmosphere or in the water can get caught in ice. Debris of all sizes and kinds can get trapped: not only seeds, plants, and ash, but vintage World War II planes. Air from the atmosphere can get caught as bubbles. The layers of ice themselves can reveal the extent of the seasons. Arctic winters are dark and cold, whereas the summers are bright and less cold. With the sun come melt, flowing water, and the detritus water brings. Summer bands in the layers are darker and messier than the ones made in winter.

Dust blown by the winds can make some layers darker than others. With so much trapped in the ice, it becomes a very precise and informative record of ancient climates.

Dansgaard's breakthrough came from applying the tools developed by Harold Urey to the Greenland ice core. Since his focus wasn't shells but ice, the work required a few modifications, but he nevertheless was able to see a climate record. He measured oxygen along an ice core over half a mile deep, representing more than 100,000 years. Dansgaard saw the remarkable chilling taking place 17,000 years ago, during the ice ages first seen by Agassiz. He also encountered a warming period 500 years ago, corresponding to when humans first settled Greenland. And he found a cooling period extending from 1700 to 1850, when much of Europe was cold and Hans Brinker was ice-skating in the canals of Amsterdam.

Dansgaard's was a rough first effort because his core, having been dug for missiles and churches, didn't allow for great scientific resolution. A scientifically useful core is drilled, sectioned, and kept in conditions that allow long stretches of unbroken ice to be analyzed. Needed were new, more precise cores. And if these data were to have meaning, he'd need to see ice from different places on the planet: from both poles and from mountaintops of different continents.

Drilling scientifically accurate cores requires collaboration among engineers, scientists, and governments working on the planet's largest ice sheets. This is expensive science: rigs need to be set up, and teams housed, in some of the most remote places on Earth. Since the 1970s a number of cores have been drilled, and to date the most complete of these are several drilled into the Greenland ice, the glaciers in Antarctica, and several mountain glaciers from around the world.

The fine-grained view of climate and ice reveals surprises. Earth's climate during the past 100,000 years has swung wildly on occasion. The ice ages weren't just long invariant cold periods:

glacial periods have witnessed warm intervals, and warm intervals have seen glacial conditions. The emerging picture is that Earth's climate depends on the heat balance of the planet—the amount of heat coming in from the sun minus the heat that escapes into space—and the ways that this heat is transferred among the oceans, land, air, and ice. Music is an analogy for what drives climate: a composition can be heard as one entity but be decomposed into rhythms, backbeats, and harmonies of different instruments acting on their own cycles. Orbital motions of the kind revealed by Milankovitch define the main cadence. The movement of heat through ocean currents, winds, and ice floes form other beats. The result of the interacting effects of these components is a system that has a long-term rhythm and short-term riffs.

Climate at the end of the last glacial period, about 12,500 years ago, exemplifies one of the riffs. At this time, when by all accounts things should have continued to warm, there was a dramatic shift to a sharp cold spell that happened in the blink of an eye in geological terms—over decades. The record from pollen, oxygen atoms, and other markers implies a climate that converted from warm to cold on a dime. Global mean temperatures changed 15 degrees in as little as a decade. If wiggles of the climate curves are like an EKG, fluctuations like this are the equivalent of planetary heart attacks. When you think of the extent to which coastlines, arable land, and deserts can be transformed by changes in global temperature of just 2 or 3 degrees, the prospect of a 15-degree shift is staggering. Yet that is the kind of change that has taken place during the history of our species.

SEEDS OF CHANGE

Orbits, climates, and ice define the way living things spread across the globe and through time. Changes in global climate fragment some populations into isolated groups separated by ice. Others are offered new migration routes, enabling them to reach portions of the globe inaccessible under previous climatic conditions. DNA of Native Americans reveals that they are derived from a single male who likely crossed the Bering Strait when an ice bridge formed during the last ice age. European populations, too, carry the signal of ice in their family trees. The DNA of many Europeans derives from populations that formerly lived in Ukraine and spread out during the last recession of ice. Ice is carried deep inside our human family tree, in the DNA we share with our diverse human cousins.

Some populations do not change; they die. The end of the last ice age in North America was a double whammy for the mammals that lived there. First, they had to deal with changing climatic conditions. On top of that, they had a new competitor and predator to deal with: people. The change in climate and the arrival of humans from Asia spelled the end for North America's saber-toothed tigers, mammoths, and ground sloths.

Still other populations change their way of life altogether.

Dorothy Garrod was known to her colleagues at Cambridge as being "cripplingly shy" and "difficult to know." Yet she was anything but shy. "My dear Jean," wrote Garrod to her cousin in 1921, "The last week in France was great fun. It was really almost too moving to be true. You crawl on your stomach for hours . . . climbing up yawning abysses (lighted only by an acetylene lamp . . .) and get knocked on the head by stalactites and on the legs by stalagmites, and in the end arrive at all sorts of wonders." Here was a woman who explored ancient worlds, experienced raw adventures, and had a lot of fun doing it. Discoverer of Neanderthal bones in caves and new archaeological sites around the globe, this "shy" woman became the first female occupant of a chaired professorship at both Oxford and Cambridge.

Digging in Shukba Cave and the surrounding fields near Jeru-

Dorothy Garrod (right) in the field.

salem, Garrod discovered odd stone tools shaped like crescents. Nothing like them had been seen before. Then she unearthed a series of mortars, grinding stones, and figurines. The people who lived there had ground wheat and practiced religion.

More digging yielded more discoveries: carefully buried dog skeletons, shelters, bodies in graves with intricate decorations, even elaborate stone sculptures. These people, whom Garrod called Natufians, had the first domesticated dogs, the first sculptures of people having sex, and elaborate burial rituals. The Natufians had settlements with hundreds of people interacting in complex societies that changed over time. Previously, human populations were nomadic: populations adapted to changing climates and food supplies by moving. Natufians exemplify novel strategies: the development of a largely sedentary culture that ranged from mobile camps to semipermanent settlements over several thousand years—ranging from fifteen thousand to eleven thousand years ago.

No population is insulated from changes to the planet, particularly the kinds of decadal climate shifts recorded in the polar ice. The Natufians lived during a period of rapid climate change about thirteen thousand years ago: a cold plunge brought glaciers to high latitudes and cold, dry, weather to lower ones. This cold snap meant that traditional grains likely became more scarce. The Natufians and their contemporaries were almost certainly stressed by this shock to the global climate system, let alone to their food supply and way of life. How did they and the cultures that followed manage?

Plump seeds, typical of domesticated plants, have been found in the remains of Natufian settlements from about eleven thousand years ago. Beginning as rare components in Natufian sites, kernels and grains become common in later human settlements. The seeds are evidence of agriculture; the mortars and pestles are signals of a society using their crops for food. With these inventions, humans no longer needed to rely on the vagaries of migrating animals for subsistence. With the development of agriculture, and more permanent settlements seen in places such as those with Natufian culture, humans could now establish institutions and cultural practices associated with stable societies.

Just as Dorothy Garrod dug in the earth to discover Natufian culture, Jonathan Pritchard, my colleague at Chicago, peers within DNA to see patterns in its structure and sequence. By comparing the DNA sequences of living humans, he can tell if our differences are due to the vagaries of chance or have been sculpted by the action of natural selection. If a particular gene offered an advantage in survival or reproduction to the people who possessed it, it should leave a signal in DNA—one that he could see using statistical techniques he developed for just this purpose. All else being equal, if selection has operated on a gene, it should be more common and less varied in a population than it would be by chance alone.

Jonathan has found stretches of human DNA that carry the signature of natural selection; these are genes that in some way affected the survival or reproduction of our ancestors. This is a kind of holy grail for biologists, because they can tell what biological traits were important. And what do these genes do? Some relate to color pigment. If the spread of human populations across the globe brought them to areas with different light levels, the genes affecting pigmentation would change, with lighter pigmentations found in populations more distant from the equator.

Other genes reflect changes to the diet. Genes that became common in some human populations relate to digesting milk, carbohydrates, and alcohol. The ability to process these products involves special enzymes that break down the characteristic sugars inside. The genes involved with these functions gained a new importance in the past ten thousand years. The ability to digest milk is evidence of the domestication of cows; processing alcohol relates to fermentation. Both are traits of agricultural and, to some degree, sedentary human communities.

The effects of rotating planets and past chills are everywhere— from the sand on the beach to exotic boulders in the landscape, even to parts of our own DNA that persist, like the tunnels of Camp Century, as artifacts of changing climates and cultures.

CHAPTER TEN

MOTHERS OF INVENTION

By 8 million years ago the shapes of the continents, oceans, and seas would be recognizable to an elementary school class today. The planet looked decidedly modern, except for one important omission: it lacked a big-brained species walking on two legs.

Hints to the revolution afoot are first seen inside rocks about 7 million years old from what are today Chad and Kenya. A French team, working at the margin of a lake bed, unearthed a chunk of a skull that has a remarkable mix of traits. With large brow ridges above the eyes and a small cranium, it looks something like a chimpanzee. But the snout and face are far too small for any chimp: these traits are decidedly humanlike. More clues come from slightly younger rocks in Kenya. The portions of femurs and other leg bones that have been found are straight, much like those of a creature that spends time moving about on two legs. Something was happening, as new kinds of apes lived, and perhaps even walked, on the planet.

These creatures certainly didn't know it, but the ground under their feet was changing. The continent of Africa was beginning to fracture. Upheavals deep within Earth caused the crust to tear, opening up a rift that began to unzip the continent from north to south. The rip started small and widened to extend about two thousand miles, from Egypt south to Mozambique. As the process continued, these rifts, like those we chased in our own hunt in Greenland's 200-million-year-old rocks, caused bulges and depressions in the surface of Earth that formed a series of valleys with mountains.

Pretty much every fossil that tells of our history 6 to 2 million years ago comes from some part of this rift system. All of them show that walking on two legs is one of our most ancient human traits, vestiges of which are seen in the creatures from Chad and Kenya. Other finds reveal a number of species with smaller faces, relatively small canine teeth, large molars in back, and a large brain, human features that arise in succession. By 1.9 million years ago, other kinds of human relatives are found outside Africa. Their bones reveal creatures fully capable of walking, even running long distances. By 200,000 years ago, members of our own species, *Homo sapiens*, were roaming Earth.

Our weather instruments for this time lie inside dust, mud, and bone. Dust trapped in the sediments gives us a hint to the levels of aridity and the direction and speed of wind. Muds from the bottom of the seas tell of the water currents and the amount of water flowing from the Nile into the sea—a measure of rainfall. The sediment at the bottom of the big rift lakes tells how the water levels fluctuated over time. The presence or absence of antelope fossils tells us whether the landscape had forests or grasslands. The presence of hippo fossils tells us whether the area was wet. Even the length of the neck of giraffe fossils tells a tale: the length of the neck can tell us if tall or short trees were around. When you know how to look, almost anything can become a thermometer, a barometer, or even an anemometer.

The level of dust in the sediments of Africa waxed and waned. All the while, antelope roamed wider, while the location of hippos and giraffes became more localized. Closed forest landscapes gave way to ones that were more open and full of grasses as the climate shifted from wet and warm to dry and cold. Despite this directionality to climate change, in any short stretch of time the environment would swing back and forth dramatically.

As any resident of a city near a mountain range knows, elevated land can exert a huge impact on the nearby climate. The development of rifts in Africa heralded a new influence on the local weather: the mountains associated with them soaked up moisture and created rain shadows. Africa's climate became fragmented between regions that were subject to torrents and those that were colder and drier.

Geographic and climatic transformation in Africa is linked to global changes. Between 2 and 3 million years ago, the ice ages began. Expanding glaciers led to lower sea levels, which, in turn, caused changes to ocean and atmospheric currents. The end result is a gradual transformation of East Africa from a land of forests to an open landscape of grasslands.

This planetary chain reaction ultimately affected our ancestors. The ability to walk on two legs that arose in forested environments in places like Chad, Ethiopia, and Kenya 7 to 4 million years ago now gained importance in the newly open savannas: our ancestors could roam large distances and use newly free hands to make tools.

The rapidly changing climates caused by changing planetary orbits and causes still unknown, characteristic of this time, also meant that any species that could adapt quickly would survive. If there is one thing a big-brained species can do, it is learn and adapt. In this crucible of environmental change—a blink of an eye in geological time—comes evidence of ever-new kinds of stone tools, shellfishing, hunting, painting, burying the dead, fire use, cooking, and Natufian agricultural societies.

CONNECTED

In the early 1980s, Stephen Stigler, a colleague of mine in the statistics department at the University of Chicago, was asked to contribute an essay to a volume celebrating one of the intellectual giants of sociology, Robert Merton. During his long career, Merton made seminal contributions to our understanding of how great ideas and innovations came to be. This effort culminated in 1957, when Merton delivered an address to a major academic society that exposed a curious pattern: as often as not, the idea we most associate with a single person was discovered by somebody else. Moreover, breakthroughs are frequently made simultaneously by different people working in different places. These are "multiples": the discovery of the same idea or invention by different people working independently.

Charles Darwin, after his trip on the *Beagle,* described the principle of natural selection for a huge tome he was preparing. Then, as his manuscript was nearing completion, he received word that Alfred Russel Wallace had independently discovered the same concept during a bout of malaria fever suffered in Indonesia a few years before. People had been looking at, drawing, and thinking about our relationship to animals for millennia, yet the discovery of one of the fundamental principles about living things happened in a virtual instant by two people working independently. Gottfried Leibniz and Sir Isaac Newton discovered calculus independently and nearly simultaneously. Elisha Gray and Alexander Graham Bell invented the telephone in the same year. And the list goes on. Many great ideas seem to come to different people at about the same time.

Virtually every graduate student or young scientist lives in fear of this situation. What is the first thought most have when they find something really great? It is not always "Whoopee!" They have promising careers to start, and they need to gain a degree of credit for their work. Their main thought is often "Who else might have already found this?"

With numerous examples in hand, Stigler coined a now famous law. In brief, when referring to a named discovery such as Hooke's law, Newtonian physics, or Darwinian theory of natural selection, he proclaimed, people should keep in mind that "no scientific discovery is named after its original discoverer." Steve Stigler named his law "Stigler's law," in a fitting tribute to its "discoverer" Merton and his disciples, who built on the work of the original "discoverer," the father of social science, Francis Bacon. The recognition of the importance of multiples in invention has itself come about many times.

The rich history of innovation is not a linear path from one person to the next but the product of a social milieu with innumerable antecedents and, as a result, multiple inventors. The actual inventor is often less important than the context in which things transpired; rather, something is "in the air" at the time of discovery. Those antecedents include the necessary conditions for inventions to happen and for the opportunities to develop and sustain them. Seeing the importance of antecedents and context, Bacon in the sixteenth century famously said, "Time is the greatest innovator."

Our bodies and genes are layer after layer of biological inventions integrated with one another over billions of years. Consequently, the biological world, just like the technological one, is loaded with multiples. The ability to breathe air, for example, has come about many times in fish, as have fins capable of moving fish about on mud and land. Lungs, or their equivalents, are seen in a number of freshwater species: some breathe with a saclike lung; others have added blood vessels to other parts of their bodies. Walking, too, is a common behavior in fish and is seen in species as different as frogfish, mudskippers, and epaulette sharks. Some fish even climb trees. Multiples are not just seen in fish; every species, even our own, reveals them in one form or another. Stigler's law applies to organs, as it does to theorems and devices.

Our own change did not happen in a vacuum: here too time was our greatest innovator. A human being could no more arise during the Devonian era, 375 million years ago, than could an iPad have been invented in the eighteenth century. It took a great number of antecedents for legs, feet, or silicon chips to come about in the first place.

What is "in the air" for the biological inventions is the state of
the air itself and its links to rock, water, and life. The bipedalism
that was so important to our species was possible only because of
changes that happened in fish, worms, and other creatures. The
shift from fins to legs happened as our ancestors went from living in water to walking on land. Transport in time to the world
of 380 million years ago, when streams and oceans were teeming
with life: fish large and small. In the water, it was a real "fish eat
fish" world at this time. Big carnivorous fish over fourteen feet
long swam in the same streams as numerous smaller armored
ones. Land at this time was not a barren ecosystem. Plants and all
sorts of invertebrate animals invaded land first, establishing lush
forests, shrublands teeming with life. The earliest fish walking
on land hit an environment that was already hospitable: loaded
with food and absent any predator. There was a push and pull
for our distant fish ancestors to move to land. Any feature that
enabled them to escape the large predators in water while capi-
talizing on the opportunities offered by the life on land would be
a distinct advantage.

Plants helped define the geography and environment for our
ancestors' shift to land. Plants have root systems that allow soils
to form and thus solid banks for the streams in which our fish
ancestors lived. The rise of plants on land, with their photosynthetic metabolisms, also is associated with enhanced levels
of oxygen in the atmosphere. Human legs owe their origins as
much to the history of trees, shrubs, and flowers as they do to
that of fish.

But fish, legs, and plants are only one stop in the story of bodies; the origin of every tissue, cell, and gene is the product of an
interrelationship between the planet and living things. Were it
not for algae and moving continents, the cellular machinery for
legs—or for any of our organs for that matter—would not even
exist. That change built on billions of years of history: from the
imbalance of matter over antimatter after the big bang to the

workings of the solar system to the recycling of Earth's crust that made our species possible. Our antecedents on Earth are as much our long lineage of animal forebears as they are the planetary and cosmic events that have been intertwined with us, and our history, since the beginning.

The American philosopher William James often said that religious experience emanates from "feeling at home in the universe." With bodies composed of particles derived from the birth of stellar bodies and containing organs shaped by the workings of planets, eroding rock, and the action of the seas, it is hard not to see home everywhere.

PAST AS PROLOGUE

The team had followed the trail of ancient rift valleys from the east coast of Greenland to the foothills of Morocco's Atlas Mountains. The layers of rock in northwest Africa took the forms of eroding sandstones and shales we were familiar with in the Arctic: trade the polar bears for goats, glaciers for small villages, and the underlying geological setting felt like home. So were the scientific goals: our success in Greenland gave us the experience and the thirst to find new places to look for ancient mammals and their tiny teeth.

As Farish and I descended to a promising valley of dusty red sandstones, our attention was pulled from the hike by the clatter of donkeys, a cue that local villagers were nearby. Typically, this kind of encounter involved a goatherd or a pack of small children. Their curiosity about us, coupled with their naturally slapstick sense of humor, enlivened many a fruitless day looking for fossils.

The approaching sound gave way to the image of two old men whose bright eyes and wide smiles belied bodies crumpled by age. The donkeys were smaller than their riders, who bore huge toothless grins, faces creased by the sun, and hands and feet gnarled and callused by years of hard labor.

The men had something to tell us, and not speaking a word of Berber, Farish and I employed the usual mix of sign language and facial expressions to convey our thoughts. It was clear that the two were trying to communicate something important, but for the life of us we couldn't follow them. Finally, in a measure of desperation, one of them took a worn, yellowed piece of paper from his robe and passed it to me. An old picture revealed it was his ID card from his days as a laborer. Then a lightbulb in my head went off: he had worked with the French paleontologists who studied these sites twenty years before, and he was trying to form a connection with us. Farish looked at the document, read the print faded by creases and years of wear, and shook his head. He exhaled softly and said, "I'm older than these guys." Farish, a fit man in his fifties, looked about forty years younger than our companions.

The impact of the planet was written over the faces, joints, and bodies of our Berber friends. Culture, technology, and economics mediate and referee our interaction with Earth. We don't need to travel to Morocco to encounter these effects; we can cross a couple of city blocks in Manhattan or Chicago to find vast differences in longevity, literacy, infant mortality, various cancers, diabetes, obesity, and heart disease.

In our past, vagaries of the planet and its geography defined the sharpest distinctions among our ancestors, whether fish, reptile, or human. But the equation is now changed, and the roots of this shift can be traced to something first seen in the rocks in Africa about 3 million years ago.

The first stone tools were used to butcher meat. Since then, we've invented devices to perform every function imaginable, from growing food to traveling under the sea. We exchange information in ever-changing ways, from alphabet to voice to digital. Our history has been one of gizmos, medicines, and technologies to make our thoughts real and expand the possibilities of our lives.

With virtually every technology and idea, our species has found new ways to insulate itself from the planet. Ever since the Natufians, stable communities that use agriculture have buffered us from the vagaries of depending on migrating animal populations for food. Clothes insulate us from changes in the weather. Tools and machines allow us to reach beyond the limitations of our physical bodies. We've even made objects that enabled us to leave the gravitational pull of the planet itself and extend our senses to other celestial bodies.

Human creativity and biology are like different instruments in an orchestra: they play separately, but together they make one score. The advent of cooking is written inside our guts and in the genes that form them. The origin of agriculture is reflected in the structure of our DNA. Our technological and cultural inventions impact our biological selves. But our biology—so defined by big brains, dexterous hands, and speaking organs—makes these inventions possible in the first place. Biology and culture have been the yin and yang of the human experience on our planet.

Are we in the process of breaking a balance that has been part of us since our beginnings in ancient savannas, forests, and caves? If we take a time machine and return to the planet one thousand years from now—or 1 million for that matter—what will control how fast we run, how long we live, or how much we learn?

Baseball statistics provide clues. We've now reached the point where the measurement of human abilities—hitting home runs, for example—needs to be subdivided between those whose performance has been enhanced by technology and those whose has not. Given new technologies that reveal how drugs can impact everything from strength to cognitive abilities, we may find that Nobel Prizes in the distant future will need the same treatment. It has been eleven thousand years since the dawn of human civilization. With the ever-increasing pace of change around us, imagine what humans will be capable of in another eleven thousand.

Seeing the impact of this technology, we can ask: Is Darwin

no longer behind the steering wheel? Has the result of millions of years of human evolution been to disconnect us from the planet and from evolution itself?

Biologists use a variety of tools to put numbers on the ideas that sprouted from Darwin. This is not idle fancy; numbers and equations help us do what science often does best: make and test specific predictions. Phrases like "survival of the fittest" simply don't cut it. To predict evolutionary change, we need real numbers describing the traits creatures possess, how they are passed on to each generation, and the ways they can impact the success of creatures in their environment. In the evolutionary sense, we define these numbers, particularly "success," as precisely as possible: in the number of viable offspring that a particular creature has over its lifetime in a particular environmental setting. If red birds have more viable offspring than green ones in one place, and color pattern has a strong genetic basis, then, all else being equal, over time natural selection could act to increase the number of red birds. Natural selection never disappears; if certain criteria are met, then it is an inevitable outcome.

Doing this kind of work on humans is complex, and the numbers we need can be inferred only to an approximation. Ideally, we'd have studies with huge sample sizes that would allow us to follow whole families to see how traits are passed on. The best records of human traits come from large clinical studies designed to follow the health of populations over long periods. The venerable Framingham Heart Study, for example, started in 1948, continues to this day and has followed about fourteen thousand people, recording births, number of children, and a variety of different traits and causes of death. Other studies have looked at arterial disease, reproductive health, and psychological factors. The largest batches of data come from countries with a civil registration system. Denmark, for one, has assembled data on 8 million people, including everything from fertility to family history.

The biologist Stephen Stearns and his colleagues crunched

these data and were led to a humbling insight. There is a very big difference among people in different parts of the world. In the developed world, with its access to medical care, food, and technological wizardry, most evolutionary pressure is on aspects of fertility: when we have offspring and how many we have. In the developing world things are very different: passing on one's genes is about mortality, particularly that of children. In one world, evolutionary success is derived from the age at which people have babies; in the other, such success is derived from survival itself. Socioeconomic, cultural, and technological differences mediate the ways evolution acts in human populations.

In our past, long-term success came from spreading one's genes and traits, often in response to changes to the environment. The major source of information passed from one generation to the next was written in DNA. The situation now is not so simple. The American scientist Norman Borlaug and his wife had three children, five grandkids, and six great-grandchildren. We can look at his entire family tree and assess the extent to which his genetic traits have been passed from generation to generation. If we transport to the future, we could assess the success of his biological traits in the gene pool: hair color, ability to curl his tongue, susceptibility to diseases, and so on. But how much will traits like those really matter for our future as a species? In addition to passing on his genes, Borlaug was the widely acclaimed father of the "green revolution," whose work on corn and wheat increased their pest resistance and yield. He is responsible for bettering or saving the lives of millions of people around the world. His ideas live in the ways others have used them and improved them and in an entire planet changed by his genius. From the people saved by agricultural and medical breakthroughs to the lives changed by great literature, philosophy, and music, the success of our species resides inside the offspring of our minds.

Like a sixty-year-old person on actuarial charts, the habitable Earth is three-quarters of the way through its calculated life expectancy. Earth is about 4.57 billion years old, and the laws of stellar physics tell of another billion years before the sun expands to the point that it bakes the possibility for life off the planet. Looking back, life got going quickly after Earth's formation—within a paltry few hundred million years. Bodies took roughly 2.5 billion years to come about. Then, one after another, heads, hands, and consciousness arose in ever more rapid succession. As in Moore's law, which famously describes the doubling power of silicon chips every twenty-four months, the biological world has witnessed exponential rates of change: it took most of the expected life span of our planet for the origin of a big-brained species using stone tools; then merely thousands for the origin of the Internet, gene cloning, and schemes of geo-engineering the atmosphere of the planet itself. Planetary and biological change have brought about a transformative moment—one in which ideas and inventions shape our bodies, the planet, and the interactions between them. Before our species hit the scene, trillions of algae took billions of years to transform the planet; now change is driven by single ideas traveling at the speed of light.

Ours is a species that can extend its biological inheritance to see vast reaches of space, know 13.7 billion years of history, and explore our deep connections to planets, galaxies, and other living things. There is something almost magical to the notion that our bodies, minds, and ideas have roots in the crust of Earth, water of the oceans, and atoms in celestial bodies. The stars in the sky and the fossils in the ground are enduring beacons that signal, though the pace of human change is ever accelerating, we are but a recent link in a network of connections as old as the heavens.

Excellent works for a general audience on the history of the universe, planet, and life grace the literature. Carl Sagan's *Cosmos* (New York: Ballantine Books, 1985), while superseded by decades of scientific discovery, remains one of the clearest and most evocative accounts of the universe and its connection to us. The story from the big bang to the formation of the planet has been told by several scientist-authors, including Lawrence Krauss, *Atom: A Single Oxygen Atom's Journey from the Big Bang to Life on Earth . . . and Beyond* (Boston: Back Bay Books, 2002); and Neil deGrasse Tyson and Donald Goldsmith, *Origins: Fourteen Billion Years of Cosmic Evolution* (New York: Norton, 2005). Richard Fortey, with his characteristic elegance, covers the history of the planet in *Earth: An Intimate History* (New York: Knopf, 2002). Fortey's book joins Tim Flannery's *Here on Earth: A Natural History of the Planet* (New York: Atlantic Monthly Press, 2011), Michael Novacek's *Terra: Our 100-Million-Year-Old Ecosystem—and the Threats That Now Put It at Risk* (New York: Farrar, Straus and Giroux, 2007), and Curt Stager's *Deep Future: The Next 100,000 Years of Life on Earth* (New York: Thomas Dunne Books, 2011) as forward-looking and richly described histories of the planet and the processes at work on it. For general and lively overviews of the history of life, see Richard Dawkins's *Ancestor's Tale: A Pilgrimage to the Dawn of Evolution* (New York: Mariner Books, 2005), Andrew Knoll's *Life on a Young Planet: The First Three Billion Years of Evolution on Earth* (Princeton, N.J.: Princeton University Press, 2004), and Brian Switek's *Written in Stone: Evolution, the Fossil Record, and Our Place in Nature* (New York: Bellevue Literary Press, 2010).

ONE   ROCKING OUR WORLD

Using the predictions of evolutionary and geological history to find fossils means employing the tools of historical geology, especially the fields of stratigraphy, sedimentology, and structural geology. Generally speaking, stratigraphy works to piece together the layers of rock in Earth to understand their ages and their relationships to one another. Sedimentology centers on elucidating the conditions that led to the formation of rocks such as the sandstones, shales, and siltstones that occasionally contain the fossils of interest to paleontologists like myself. Were the rocks originally deposited by the action of lakes, streams, or oceans, or by some other earthly process? Structural geology seeks to make sense of the movements and forces at work on the rocks relative to one another during the millions of years from their deposition as sediments to their presence as layers today. General references on these fields abound. For a captivating primer that requires absolutely no prior knowledge, see Marcia Bjornerud, *Reading the Rocks: The Autobiography of the Earth* (New York: Basic Books, 2005). Fortey's *Earth* also fits into this category. See also Walter Alvarez's excellent *The Mountains of St. Francis: Discovering the Geological Events That Shaped the Earth* (New York: Norton, 2008).

Bill Amaral's insight into the fossil-bearing potential of the Triassic-age rocks in Greenland began with the *Shell Oil Guide to the Permian and Triassic of the World*. The library discard saved from the trash was K. Perch-Nielsen et al., "Revision of Triassic Stratigraphy of the Scoresby Land and Jameson Land Region, East Greenland," *Meddelelser om Grønland* 193 (1974): 94–141. This reference ultimately led Bill, Chuck, and Farish to the elegant sedimentological work of Lars Clemmensen, a Danish sedimentologist. These papers— L. B. Clemmensen, "Triassic Lithostratigraphy of East Greenland Between Scoresby Sund and Kejser Franz Josephs Fjord," *Grønlands geologiske undersøgelse* (1980), and L. B. Clemmensen, "Triassic Rift Sedimentation and Palaeogeography of Central East Greenland," *Geological Survey of Greenland, Bulletin*, no. 136 (1980): 5–72—became a kind of Rosetta stone because they revealed the fossil potential of the rocks and their similarity to those of eastern North America (described in P. E. Olsen, "Stratigraphic Record of the Early Mesozoic Breakup of Pangea in the Laurasia-Gondwana Rift System," *Annual Reviews of Earth and Planetary Science* 25 [1997]: 337–401). This was a eureka moment in the library.

Coming to grips with the past inside the rocks is as much about practical matters—food, boot selection, and learning to see—as about great ideas at stake. The first of these considerations is derived from one of Napoléon's insights: armies run on their stomachs. You can have the best scientific preparation on the planet, but if the food is terrible, things can go awry very quickly. When a field crew eats well and meals become an event to anticipate, folks can endure privations of weather, boredom, and the drudgery of failure that a new fossil hunt brings. Long days spent wet and cold, finding nothing, can be rescued by comfort foods awaiting people when they return to the tent at night. Before departing for the Arctic, we prepare by dehydrating many of our own vegetables and fruits to devise menus that have a diversity of tastes, textures, and smells. Come to my lab in April before a field season and you might smell kiwis, strawberries, or San Marzano tomatoes in the dehydrator. We even bake bread in the field, knowing that the smell of a rising loaf can not only sell a house but also soothe a surly field crew. In the field, the bread tastes like the finest French baguette. Unfortunately, our creations, having the consistency and density more of a building material than of an edible substance, would be an insult if served at home.

We knew few of these tricks the first season in 1988. The meals were all pre-packaged dehydrated affairs with fancy labels to make one salivate and sumptuous names such as veal scallopini, chicken marsala, and turkey tetrazzini. After two weeks of gorging on these bags in the field, we noticed that they all tasted the same. A depressing confirmation came when I read the ingredients lists: all our fancy dinners were essentially the same meal with a different label, in a different-colored bag, with a different-shaped pasta. Revealing this discovery to my colleagues did not help matters; the ensuing run on hot sauce and spices left us with no way to vary tastes. Needless to say, I lost a lot of weight that year.

Our recipes for dehydrated meals can be found at http://tiktaalik.uchicago.edu. They do the trick after a day of slogging through tundra or clambering over scree, use a minimum of fuel and water during preparation, can be modified for everyone from a vegan to a ravenous carnivore, and aren't heavy to pack or ship. You can even serve the meals at home to company you never wish to see again.

For a good nontechnical introduction to the geological history of eastern North America, see Chet Raymo and Maureen E. Raymo, *Written in Stone* (Hensonville, N.Y.: Black Dome Press, 2007). The story of Lull's dinosaur in the bridge abutment can be found in Edwin H. Colbert, *Men and Dinosaurs* (New York: E. P. Dutton, 1968).

The Greenland discoveries are described in F. A. Jenkins Jr. et al., "A Late Triassic Continental Fauna from the Fleming Fjord Formation, Jameson Land, East Greenland," in *The Nonmarine Triassic*, ed. S. G. Lucas and M. Morales (Albuquerque: New Mexico Museum of Natural History and Science, 1993), 74; F. A. Jenkins Jr. et al., "A New Record of Late Triassic Mammals from the Fleming Fjord Formation, Jameson Land, East Greenland," in Lucas and Morales, *Nonmarine Triassic*, 94. The most important of the mammals we found is described in F. A. Jenkins Jr. et al., "Haramiyids and Triassic Mammalian Evolution," *Nature* 385 (1997): 715–18.

General references on the origin of mammals and the relevance of little teeth to our own branch of the evolutionary tree can be found in Zofia Kielan-Jaworowska, Richard L. Cifelli, and Zhe-Xi Luo, *Mammals from the Age of Dinosaurs: Origins, Evolution, and Structure* (New York: Columbia University Press, 2004); and Z.-X. Luo, "Commentary on Mammalian Dental Evolutionary Development," *Nature* 465 (2010): 669.

TWO   BLASTS FROM THE PAST

The relative numbers of the atoms in the human body were taken from Robert W. Sterner and James J. Elser, *Ecological Stoichiometry: The Biology of Elements from Molecules to the Biosphere* (Princeton, N.J.: Princeton University Press, 2002), chap. 1. This is of course not a true chemical formula, as the ratios of elements in us compose not a single unique molecule, like a crystal of salt, but a body consisting of numerous different kinds of them.

The notion of a tree of life that connects all creatures living and extinct is one of the insights of the Darwinian revolution. It makes specific and testable predictions and allows us to formulate hypotheses. Background discussion of these methods written for a general audience is found in Dawkins's *Ancestor's Tale*. For those who wish a treatment written for practitioners, try E. O. Wiley et al., *The Compleat Cladist: A Primer of Phylogenetic Procedures*, special publication no. 19 (Lawrence: University of Kansas, Museum of Natural History, 1991), http://www.archive.org/stream/compleatcladistp00wile#page/ n5/mode/2up. To get a real taste for the discipline with its applications and vigorous debates, try some of the journals in the field: *Cladistics* and *Systematic Biology*.

The story of Henrietta Leavitt's work is published in E. C. Pickering, "Periods of 25 Variable Stars in the Small Magellanic Cloud," *Harvard College Observatory Circular* 173 (1912): 1–3. The story of her work, and that of the other women in the observatory, is in Nina Byers and Gary Williams, eds., *Out of Shadows: Contributions of Twentieth-Century Women to Physics* (New York: Cambridge University Press, 2006); and Jacob Darwin Hamblin, *Science in the Early Twentieth Century: An Encyclopedia* (Santa Barbara, Calif.: ABC-CLIO, 2005), 181–84.

General discussion of the big bang and its consequences can be found in Krauss, *Atom;* Tyson and Goldsmith, *Origins;* Simon Singh, *Big Bang: The Origin of the Universe* (New York: HarperCollins, 2005); and Steven Weinberg, *The First Three Minutes,* updated ed. (New York: Basic Books, 1993).

Operation Ivy Mike, the first test of the Teller–Ulam device, is described in Richard Rhodes, *Dark Sun: The Making of the Hydrogen Bomb* (New York: Simon & Schuster, 1995).

THREE   LUCKY STARS

In the days since the "nebular hypothesis" of Swedenborg, Kant, and Laplace, the origin of the planets of the solar system has been an active area of research, discovery, and debate. General background can be found in Tyson and Goldsmith, *Origins*. A general review of the dynamics of the formation of Earth is found in R. M. Canup, "Accretion of the Earth," *Philosophical Trans-*

actions of the Royal Society A 366 (2008): 4061–75. For those with a quantitative background who want to immerse themselves in the field by reading original scientific papers, go to the main scientific journal of the field, *Icarus: The International Journal of Solar System Studies,* the official publication of the Division of Planetary Sciences of the American Astronomical Society.

Harry McSween has written wonderful books on the solar system, meteoritics, and cosmochemistry. See in particular *Stardust to Planets: A Geological Tour of the Solar System* (New York: St. Martin's Press, 1993). For an excellent review of the dynamics of the solar system, see J. Kelly Beatty, Carolyn C. Petersen, and Andrew Chaikin, *The New Solar System,* 4th ed. (Cambridge, Mass.: Sky Publishing, 1999).

The field of cosmochemistry is concerned with the chemical analysis of meteors, lunar rocks, and other extraterrestrial materials. On November 29, 2011, the *Proceedings of the National Academy of Sciences* (*PNAS*) ran a special issue with a number of excellent reviews. The opening review is a useful overview of the field and the special issue: G. MacPherson and M. H. Thiemens, "Cosmochemistry: Understanding the Solar System Through Analysis of Extraterrestrial Materials," *PNAS* 108 (2011): 19130–34.

Studies of the age of Earth have, themselves, a rich history. A source now several decades old but a resource for history, detail, and method, is G. Brent Dalrymple, *The Age of the Earth* (Stanford, Calif.: Stanford University Press, 1991). For a more recent treatment, see G. B. Dalrymple, "The Age of the Earth in the Twentieth Century: A Problem (Mostly) Solved," in *The Age of the Earth: From 4004 B.C. to A.D. 2002,* Geological Society, London, Special Publication 190, ed. C. L. E. Lewis and S. J. Knell (London: Geological Society, 2001), 205–21. This special publication from the London Geological Society contains a veritable feast of papers on the age of Earth, the history of this field of study, and the methods used.

Zircons are magnificent windows into early Earth. See J. W. Valley, W. H. Peck, and E. M. King, "Zircons Are Forever," *Outcrop,* University of Wisconsin-Madison Geology Alumni Newsletter (1999), 34–35. For a scientific paper on this, see S. A. Wilde et al., "Evidence from Detrital Zircons for the Existence of Continental Crust and Oceans on the Earth 4.4 Gyr Ago," *Nature* 409 (2001): 175–78. For a general treatise on the hunt for and meaning of the oldest rocks on the planet, see Martin Van Kranendonk, R. Hugh Smithies, and Vickie C. Bennett, eds., *Earth's Oldest Rocks* (Boston: Elsevier, 2007). That volume has an enormous amount of information written by and for specialists in the field.

Geologists tell time in rocks in two ways: relatively and absolutely. See Doug Macdougall, *Nature's Clocks: How Scientists Measure the Age of Almost Everything* (Berkeley: University of California Press, 2008). Relative time is a matter of understanding the relationships between different layers—layers above are generally younger than those below. The situation gets challenging when the layers are highly altered. Piecing together the layer history comes from understanding the faults and movements that shifted rocks and layers about.

Calculating absolute time in rocks and minerals depends on understanding the radioactive decay of atoms. Some atoms have an unstable configuration of electrons, neutrons, and protons, causing them to lose or gain components. As they do this, their atomic weights can change, and they become new forms. The important point is that this transformation happens at rates that are physical constants, known as the half-life. The half-life of an atom is the time required for one-half of a sample to decay, or transform, into its daughters. If you know the amounts of parent atoms, daughter atoms, and the half-life, then you can calculate the time that the atoms have been decaying. A number of atoms are useful to geologists: uranium 238, argon 39, and carbon 14, for example. In general, you try to match atoms for the job: atoms with the slowest decay rates are useful for the oldest rocks, whereas those that decay faster are useful for more recent ones. Uranium 238, with its long half-life, is useful for questions about the most ancient phases of Earth. Carbon 14 has such a rapid decay rate it is useful for more recent events, such as those of human history and culture.

In zircons, such as those from the Jack Hills, the isotopes (atomic versions) of uranium and lead are most useful. Uranium 238 decays into a stable daughter isotope, lead 206, with a half-life of 4.5 billion years. When uranium was incorporated in the zircon when it was formed, the clock started ticking: the slow transformation to lead 206 began. Looking at that zircon now, we make the reasonable assumption that all of the lead 206 has come from the decay of uranium. Knowing the ratios of parent and daughter isotopes and the half-life allows the age of the zircon to be calculated.

A general timescale for the major events in the history of the solar system and Earth is in F. Albarede, "Volatile Accretion History of the Terrestrial Planets and Dynamic Implications," *Nature* 461 (2009): 1227–33.

We've witnessed a number of different ideas about where the water of the planet came from. For a long time, it was thought that the main source was icy comets. That hypothesis was challenged when cometary water was sampled during a satellite visit to Hale-Bopp as it came close to Earth. These measurements revealed that the comet's water had a different chemical signature than the water of Earth's oceans. The comet hypothesis was in trouble until measurements were taken more recently of another comet: this one, Hartley 2, has more Earth-like water. Now a number of potential sources exist, and none are mutually exclusive: comets, asteroids, even squeezed or condensed from the constituents of early Earth. Reviews of the evidence are in N. H. de Leeuw et al., "Where on Earth Has Our Water Come From?," *Chemical Communications* 46 (2010): 8923–25; M. J. Drake and H. Campins, "Origin of Water in the Terrestrial Planets," *Proceedings of the International Astronomical Union* 1, no. S229 (2006): 381–94. The discovery of ocean-like water on a Kuiper belt comet, Hartley-2, is described in P. Hartogh et al., "Ocean-Like Water in the Jupiter-Family Comet 103P/Hartley 2," *Nature* 478 (2011): 218–20. For images of potential water in the polar craters of Mercury, see NASA's website: http://www.nasa.gov/mission_pages/messenger/multimedia/messenger _orbit_image20120322_3.html.

For the formation of the different planets of the solar system, and the relationships among them, see R. M. Canup, "Origin of Terrestrial Planets and the Earth-Moon System," *Physics Today*, April 2004, 56–62.

FOUR ABOUT TIME

The origin of the moon has been the subject of a large number of scientific papers in recent years. For a sampling, with references, see R. M. Canup, "Formation of the Moon," *Annual Review of Astronomy and Astrophysics* 42 (2004): 441–75; R. M. Canup and K. Righter, eds., *Origin of the Earth and Moon* (Tucson: University of Arizona Press, 2000); Canup, "Origin of Terrestrial Planets and the Earth-Moon System."

For a history of the ways we keep time, see the work of Anthony Aveni, in particular *Empires of Time: Calendars, Clocks, and Cultures* (Boulder: University of Colorado Press, 2002).

The idea that clocks are embedded throughout the natural world is explored in detail in Macdougall, *Nature's Clocks*.

A wonderful book on clocks, time, and our perception of time is Robert Levine, *A Geography of Time: On Tempo, Culture, and the Pace of Life* (New York: Basic Books, 1998). Michel Siffre's cave experience is documented in his personal account, *Beyond Time* (New York: McGraw-Hill, 1964).

The story of Curt Richter's life's work can be found in his biographical memoir published by the National Academy of Sciences, in *Biographical Memoirs*, vol. 65 (Washington, D.C.: National Academy Press, 1994), http://www.nap .edu/catalog.php?record_id=4548.

The story of Seymour Benzer and the discovery of the molecular basis for circadian clocks, among other things, is in Jonathan Weiner's wonderful *Time, Love, and Memory: A Great Biologist and His Quest for the Origins of Behavior* (New York: Vintage, 2000).

The starting point for learning more about biological clocks is John D. Palmer's readable and often funny *The Living Clock* (Oxford: Oxford University Press, 2002). If you want more detail from the primary literature itself, then proceed to the papers in the following paragraphs.

The Benzer lab's discovery of clock mutants is detailed in R. J. Konopka and S. Benzer, "Clock Mutants of *Drosophila melanogaster*," *PNAS* 68 (1971): 2112–16. The trio of labs that cloned the gene and explored its biological ramifications were Jeffrey Hall's (Brandeis), Michael Rosbash's (Brandeis), and

Michael Young's (Rockefeller). The biology of these circadian clock mutants in diverse creatures is discussed in a number of papers, including Z. S. Sun et al., "RIGUI, a Putative Mammalian Ortholog of the Drosophila Period Gene," *Cell* 90 (1997): 1003–11; H. Tei et al., "Circadian Oscillation of a Mammalian Homologue of the *Drosophila* Period Gene," *Nature* 389 (1997): 512–16; M. W. Young and S. A. Kay, "Time Zones: A Comparative Genetics of Circadian Clocks," *Nature Reviews Genetics* 2 (2001): 702–15; W. Yu and P. E. Hardin, "Circadian Oscillators of *Drosophila* and Mammals," *Journal of Cell Science* 119 (2006): 4793–95; E. E. Hamilton and S. A. Kay, "Snap-Shot: Circadian Clock Proteins," *Cell* 135 (2008); K. Lee, J. J. Loros, and J. C. Dunlap, "Interconnected Feedback Loops in the *Neurospora* Circadian System," *Science* 289 (2000): 107–10; E. Tauber et al., "Clock Gene Evolution and Functional Divergence," *Journal of Biological Rhythms* 19 (2004): 445–58; D. Bell-Pedersen et al., "Circadian Rhythms from Multiple Oscillators: Lessons from Diverse Organisms," *Nature Reviews Genetics* 6 (2005): 544–56.

The circadian clock and its evolution have been the subjects of a number of excellent reviews and books in the scientific literature. An entrée to this body of work can be found in the following papers: J. Dunlap, "Molecular Basis for Circadian Clocks," *Cell* 96 (1999): 271–90; M. Rosbash, "Implications of Multiple Circadian Clock Origins," *PLoS Biology* 7 (2009): 17–25; S. Panda, J. B. Hogenesch, and S. A. Kay, "Circadian Rhythms from Flies to Human," *Nature* 417 (2002): 329–35.

The similarity of sleep mechanisms in diverse organisms is discussed in detail in C. Cirelli, "The Genetic and Molecular Regulation of Sleep: From Fruit Flies to Humans," *Nature Reviews Neuroscience* 10 (2009): 549–60; and Panda et al., "Circadian Rhythms from Flies to Human."

The use of the shells of invertebrate animals as timepieces to measure the length of days over time is found in Z. Zhenyu et al., "The Periodic Growth Increments of Biological Shells and the Orbital Parameters of Earth-Moon System," *Environmental Geology* 51 (2006): 1271–77. The evolution of circadian clocks is discussed in D. A. Paranjpe and V. K. Sharma, "Evolution of Temporal Order in Living Organisms," *Journal of Circadian Rhythms* 3 (2005): 7–17.

A general reference on sleep medicine is Meir H. Kryger, Thomas Roth, and William C. Dement, *Principles and Practice of Sleep Medicine* (Philadelphia: Saunders, 2005). A short but pithy review of the relationships between clocks and clinical conditions is in A. R. Barnard and P. M. Nolan, "When Clocks Go Bad: Neurobehavioural Consequences of Disrupted Circadian Timing," *PLoS Genetics* 4 (2008).

The relationship of DNA replication, circadian clocks, and cancer is discussed in S. Mitra, "Does Evening Sun Increase the Risk of Skin Cancer?," *PNAS* 108, no. 47 (2011): 18857–58; S. Gaddameedhi et al., "Control of Skin Cancer by the Circadian Rhythm," *PNAS* 108 (2011): 18790–95; and S. Sahar and P. Sassone-Corsi, "Metabolism and Cancer: The Circadian Clock Connection," *Nature Reviews Cancer* 9 (2009): 886–96.

The full story of the Hindostan grave markers is discussed in E. Kvale et al., "The Art, History, and Geoscience of Hindostan Whetstone Gravestones in Indiana," *Journal of Geoscience Education* 48 (2000): 337–42. More information on the kind of rock that the stones are taken from, called rhythmites, is in B. W. Flemming and A. Bartholomä, *Tidal Signatures in Modern and Ancient Sediments* (Oxford: Blackwell Science, 1995).

FIVE THE ASCENT OF BIG

The discovery of the earliest living things, and Barghoorn's life, are discussed in Elso Barghoorn's biographical memoir, published by the National Academy of Sciences, in *Biographical Memoirs,* vol. 87 (Washington, D.C.: National Academy Press, 2005), http://www.nap.edu/html/catalog.php?record_id=11522.

A lively personal account of discovery of early life is written by one of Barghoorn's students, now an eminence in his own right, J. William Schopf, *Cradle of Life* (Princeton, N.J.: Princeton University Press, 2001). A similarly excellent account is Martin Brasier, *Darwin's Lost World: The Hidden History of Animal Life* (Oxford: Oxford University Press, 2009). The recognition of early fossils as the remnants of living things is challenging and often spawns debate, as it has between Schopf and Brasier. A review, and one side of these debates, is in M. D. Brasier et al., "Earth's Oldest (c. 3.5Ga) Fossils and the 'Early Eden Hypothesis': Questioning the Evidence," *Origins of Life and Evolution of the Biosphere* 34 (2004): 257–60. At the time of this writing, the oldest fossil evidence for life is either Schopf's, described in his book, or Brasier's, found in D. Wacey et al., "Microfossils of Sulphur-Metabolizing Cells in 3.4-Billion-Year-Old Rocks of Western Australia," *Nature Geoscience* (2011), doi:10.1038/ngeo1238.

Galileo's discussion of size and gravity is in the definitive and readable text with translation by Stillman Drake and commentary by Stephen Jay Gould, Albert Einstein, and J. L. Heilbron, *Dialogue Concerning the Two Chief World Systems: Ptolemaic and Copernican* (New York: Modern Library, 2001).

The stories of van Leeuwenhoek were taken from Clifford Dobell, ed., *Antony van Leeuwenhoek and His "Little Animals"* (New York: Dover, 1960). The description of his likely microscope is reprinted in Clair L. Stong, *The "Scientific American" Book of Projects for the Amateur Scientist* (New York: Simon & Schuster, 1960).

Algae and the rise of oxygen are discussed, and well referenced, in Andrew Knoll's *Life on a Young Planet,* cited above. The importance of oxygen for the evolution and history of life is discussed in two other engaging books: Nick Lane, *Oxygen: The Molecule That Made the World* (Oxford: Oxford University

Press, 2003); and Peter D. Ward, *Out of Thin Air: Dinosaurs, Birds, and Earth's Ancient Atmosphere* (Washington, D.C.: Joseph Henry Press, 2006). A general review of oxygen over geological time (looking at more recent events) is in R. A. Berner et al., "Phanerozoic Atmospheric Oxygen," *Annual Review of Earth and Planetary Sciences* 31 (2003): 105–34.

The timing of the rise of oxygen, called the Great Oxygenation Event, appears as not a single increase but several scattered over hundreds of millions of years. Scientific papers include L. R. Kump, "The Rise of Atmospheric Oxygen," *Nature* 451 (2007): 277–78; A. Bekker et al., "Dating the Rise of Atmospheric Oxygen," *Nature* 427 (2004): 117–20; and H. Holland, "The Oxygenation of the Atmosphere and Oceans," *Philosophical Transactions of the Royal Society B* 361 (2006): 903–15.

Befitting such a complicated situation as the rise of oxygen, there are a number of different hypotheses for its cause. The one discussed here is outlined in L. R. Kump and M. E. Barley, "Increased Subaerial Volcanism and the Rise of Atmospheric Oxygen 2.5 Billion Years Ago," *Nature* 448 (2007): 1033–37.

The Darlington-Barbour "dropping the frogs off the roof of the MCZ" story was the stuff of legend when I was a graduate student there in the 1980s. That, and the adventure with the crocodile, are discussed in the National Academy of Sciences' biographical memoir of Philip Darlington written by his Harvard colleague E. O. Wilson, in *Biographical Memoirs*, vol. 60 (Washington, D.C.: National Academy Press, 1991), http://www.nap.edu/catalog.php?record_id=6061.

J. B. S. Haldane's essay "On Being the Right Size" was originally published in 1926 and is available at http://www.physlink.com/Education/essay_haldane.cfm.

The study of the relationship between size and other biological features is known as allometry. The literature in the field is vast, but a number of reviews can get one up to speed pretty rapidly. After Haldane's paper, there is the seminal paper of Stephen Jay Gould, produced when he was a college student. It remains an important contribution over forty-five years later: "Allometry and Size in Ontogeny and Phylogeny," *Biological Reviews of the Cambridge Philosophical Society* 41 (1966): 587–638. The history of the concept is in J. Gayon, "History of the Concept of Allometry," *American Zoologist* 40 (2000): 748–58. A general book in the field, with extensive references, is William A. Calder, *Size, Function, and Life History* (Mineola, N.Y.: Dover, 1996). The eminent biologist John Tyler Bonner has written an outstanding volume for general audiences on the consequences of size: *Why Size Matters: From Bacteria to Blue Whales*, rev. ed. (Princeton, N.J.: Princeton University Press, 2011).

The challenges for small creatures moving about in fluids are captured in E. M. Purcell, "Life at Low Reynolds Number," *American Journal of Physics*

45 (1977): 3–11. Went's musings on the importance of size for our abilities are in F. W. Went, "The Size of Man," *American Scientist* 56 (1968): 400–413.

Preston Cloud's worldview, written for a general audience, is in his *Cosmos, Earth, and Man: A Short History of the Universe* (New Haven, Conn.: Yale University Press, 1980). His National Academy biographical memoir holds the trajectory of this career and anecdotes discussed herein; see *Biographical Memoirs,* vol. 67 (Washington, D.C.: National Academy Press, 1995).

A general review, with lots of references, of the factors that control size in flies is in S. Oldham et al., "Genetic Control of Size in *Drosophila,*" *Philosophical Transactions of the Royal Society B* 355 (2000): 945–52.

A review of size-control genes and their similarity in flies and people is in J. Dong et al., "Elucidation of a Universal Size-Control Mechanism in *Drosophila* and Mammals," *Cell* 130 (2007): 1120–33.

The cost of size, particularly that resulting from living in an oxygen-rich environment, is outlined in Q. Zeng and W. Hong, "The Emerging Role of the Hippo Pathway in Cell Contact Inhibition, Organ Size Control, and Cancer Development in Mammals," *Cancer Cell* 13 (2008): 188–92; D. Pan, "The Hippo Signaling Pathway in Development and Cancer," *Developmental Cell* 19, no. 4 (2010): 491–505; and C. Badouel, A. Garg, and H. McNeill, "Herding Hippos: Regulating Growth in Flies and Man," *Current Opinion in Cell Biology* 21, no. 6 (2009): 837–43.

SIX   CONNECTING THE DOTS

The theory of plate tectonics was sprung by a network of scientists around the globe. There are several excellent resources on the history of the theories of continental drift and plate tectonics, notable among them Naomi Oreskes and H. E. Le Grand, eds., *Plate Tectonics: An Insider's History of the Modern Theory of the Earth* (Boulder, Colo.: Westview Press, 2003); Naomi Oreskes, *The Rejection of Continental Drift: Theory and Method in American Earth Science* (New York: Oxford University Press, 1999); and David M. Lawrence, *Upheaval from the Abyss: Ocean Floor Mapping and the Earth Science Revolution* (New Brunswick, N.J.: Rutgers University Press, 2002).

Eduard Suess's quotation and life story are taken from his obituary, written by an American eminence in earth science, Charles Schuchert of Yale, *Science,* June 26, 1914, 933–35.

Alfred Wegener's life, work, and impact are discussed in Roger M. McCoy, *Ending in Ice* (Oxford: Oxford University Press, 2006).

The oral history project in which Marie Tharp talks of her work is at http://www.aip.org/history/ohilist/22896_1.html.

One of the classics, describing the accumulation of evidence in support of the theory, is a short book, written over thirty years ago, that is well worth a read to expand the treatment in this chapter: Seiya Uyeda, *The New View of the Earth* (San Francisco: W. H. Freeman, 1978). Consult also Philip Kearey and Frederick J. Vine, *Global Tectonics* (London: Blackwell Science, 1996).

Frederick Vine's paper is F. J. Vine and D. H. Matthews, "Magnetic Anomalies over Oceanic Ridges," *Nature* 199 (1963): 947–49.

A short biography of John Tuzo Wilson is at http://gsahist.org/gsat/gt01sept24_25.htm. For insights into the discoveries discussed in this chapter, see J. T. Wilson, "A Revolution in Earth Science," *Geotimes* 13 (1968): 10–16; and J. T. Wilson, "Did the Atlantic Close and Then Re-open?" *Nature* 211 (1966): 676–81. You can watch Wilson describe his theories of faults at http://www.youtube.com/watch?v=OmrXy65O6fY as well as his approach to science at http://www.youtube.com/watch?v=fdSwEFyurDY.

The link between mammalian biology (placentation, size, and metabolism) and tectonic change is found in the work of Paul Falkowski and his colleagues: P. Falkowski et al., "The Rise of Oxygen over the Past 205 Million Years and the Evolution of Large Placental Mammals," *Science* 309 (2005): 2202–4. For other proposed effects of oxygen on the history of life, see also Ward, *Out of Thin Air*, and Berner et al., "Phanerozoic Atmospheric Oxygen," for perspective on the importance of oxygen and its links to other planetary processes.

SEVEN   KINGS OF THE HILL

William Smith, his map, and his struggles are the topic of Simon Winchester's *Map That Changed the World* (New York: Viking, 2001). John Phillips's ideas and work are in Jack Morrell, *John Phillips and the Business of Victorian Science* (London: Ashgate, 2005). See Phillips's *Treatise on Geology* (Surrey: Ashgate Media, 2001) for a firsthand account of his views.

For one-stop intellectual shopping on the ideas and writings of Cuvier, see M. J. S. Rudwick, *Georges Cuvier, Fossil Bones, and Geological Catastrophes: New Translations and Interpretations of the Primary Texts* (Chicago: University of Chicago Press, 1999).

Histories of the concept of extinction are found in a number of highly readable accounts, including Walter Alvarez, *T. rex and the Crater of Doom* (New York: Vintage, 1999); David Sepkoski and Michael Ruse, eds., *The Paleobiological Revolution* (Chicago: University of Chicago Press, 2009); and M. J. S. Rudwick, *The Meaning of Fossils: Episodes in the History of Paleontology* (Chicago: University of Chicago Press, 1985).

Details of Norman Newell's career were taken from his obituary in the *Journal of Paleontology* 80 (2006): 607–8. Papers of his relevant to extinction include "Crises in the History of Life," *Scientific American* 208 (1963): 76–92;

and "Mass Extinctions at the End of the Cretaceous Period," *Science* 149 (1965): 922–24.

The fifty volumes of the *Treatise on Invertebrate Paleontology* remain available at the University of Kansas Paleontological Institute (http://paleo.ku .edu/pdf/brochure.pdf).

Newell was one of several calling out loud for the reality of mass extinction. Another passionate voice was Otto Schindewolf: see his "Über die möglichen Ursachen der grossen erdgeschichtlichen Faunenschnitte," *Neues Jahrbuch für Geologie und Paläontologie. Monatshefte* (1954): 457–65.

The impact hypothesis for the end-Cretaceous event is discussed in Alvarez's book written for a general audience, *T. rex and the Crater of Doom*. The original scientific paper describing the impact hypothesis is L. W. Alvarez et al., "Extraterrestrial Cause for the Cretaceous-Tertiary Boundary Extinction," *Science* 208 (1980): 1095–108.

The other mass extinctions in the fossil record do not appear to be caused by impacts. For detailed discussion of these other events, see Michael J. Benton, *When Life Nearly Died: The Greatest Mass Extinction of All Time* (New York: Thames & Hudson, 2003); Douglas H. Erwin, *Extinction: How Life on Earth Nearly Ended 250 Million Years Ago* (Princeton, N.J.: Princeton University Press, 2008); George R. McGhee, *The Late Devonian Mass Extinction* (New York: Columbia University Press, 1996); David M. Raup, *The Nemesis Affair: A Story of the Death of Dinosaurs and the Ways of Science* (New York: Norton, 1999); and Peter D. Ward, *Rivers in Time* (New York: Columbia University Press, 2002).

The meeting at Woods Hole, along with the interactions of Schopf, Raup, Gould, and Sepkoski, is in Sepkoski and Ruse, *Paleobiological Revolution*.

Sepkoski's database is J. John Sepkoski Jr., *A Compendium of Fossil Marine Animal Genera*, Bulletins of American Paleontology, 364 (Ithaca, N.Y.: Paleontological Research Institution, 2002), http://strata.geology.wisc.edu/jack/.

His database revealed major patterns in the history of life in the oceans. These insights are detailed in J. J. Sepkoski Jr., "Patterns of Phanerozoic Extinction: A Perspective from Global Data Bases," in *Global Events and Event Stratigraphy*, ed. O. H. Walliser (Berlin: Springer, 1996), 35–51; D. M. Raup and J. J. Sepkoski Jr., "Mass Extinctions in the Marine Fossil Record," *Science* 215 (1995): 1501–3; D. M. Raup and J. J. Sepkoski Jr., "Periodicity of Extinctions in the Geologic Past," *PNAS* 81 (1984): 801–5.

The work of David Jablonski was the subject of a wonderful piece by David Quammen, "The Weeds Shall Inherit the Earth," *Independent* (London), November 22, 1998, 30–39. Original papers of Jablonski's used in this chapter include D. Jablonski, "Heritability at the Species Level: Analysis of Geographic Ranges of Cretaceous Mollusks," *Science* 238 (1987): 360–63; D. Jablonski and G. Hunt, "Larval Ecology, Geographic Range, and Species

Survivorship in Cretaceous Mollusks: Organismic vs. Species-Level Explanations," *American Naturalist* 168 (2006): 556–64; D. Jablonski, "Extinction and the Spatial Dynamics of Biodiversity," *PNAS* 105, no. S1 (2008): 11528–35; and D. Jablonski, "Lessons from the Past: Evolutionary Impacts of Mass Extinctions," *PNAS* 98 (2001): 5393–98.

The correlation of a rise in mammal diversity with the ecological vacuum produced by the end-Cretaceous extinction is supported most recently in R. W. Meredith et al., "Impacts of the Cretaceous Terrestrial Revolution and KPg Extinction on Mammal Diversification," *Science* 334 (2010): 521–24.

EIGHT FEVERS AND CHILLS

Paul Tudge's memorable flight over the Arctic was described in the original news accounts in 1986; see M. Lemonick, C. Tower, and D. Webster, "Unearthing a Fossil Forest," *Time*, September 22, 1986. Original papers from the primary literature include J. F. Basinger, "Early Tertiary Floristics and Paleoclimate in the Very High Latitudes," *American Journal of Botany* 76, no. S6 (1989): 158; J. F. Basinger, "The Fossil Forests of the Buchanan Lake Formation (Early Tertiary), Axel Heiberg Island, Canadian Arctic Archipelago: Preliminary Floristics and Paleoclimate," in *Tertiary Fossil Forests of the Geodetic Hills, Axel Heiberg Island, Arctic Archipelago*, Geological Survey of Canada Bulletin no. 403, ed. R. L. Christie and N. J. McMillan (Ottawa: Geological Survey of Canada, 1991), 39–65; D. R. Greenwood and J. F. Basinger, "The Paleoecology of High-Latitude Eocene Swamp Forests from Axel Heiberg Island, Canadian High Arctic," *Review of Palaeobotany and Palynology* 81, no. 1 (1994): 83–97; D. R. Greenwood and J. F. Basinger, "Stratigraphy and Floristics of Eocene Swamp Forests from Axel Heiberg Island, Canadian Arctic Archipelago," *Canadian Journal of Earth Sciences* 30, no. 9 (1992): 1914–23; B. A. Lepage and J. F. Basinger, "Early Tertiary Larix from the Buchanan Lake Formation, Canadian Arctic Archipelago, and a Consideration of the Phytogeography of the Genus," in Christie and McMillan, *Tertiary Fossil Forests of the Geodetic Hills*, 67–82.

Colbert's discovery in Antarctica was also the subject of news accounts of the time; see "New Life for Gondwanaland," *Time*, March 22, 1968. You can hear Colbert himself tell you of his Antarctica work at http://www.youtube .com/watch?v=UNe5SGkQP7Q. The discovery of *Lystrosaurus* from Antarctica is described in E. Colbert, "*Lystrosaurus* from Antarctica," *American Museum Novitates* 2535 (1974): 1–44, http://digitallibrary.amnh.org/dspace/ bitstream/handle/2246/5462//v2/dspace/ingest/pdfSource/nov/N2535 .pdf?sequence=1.

The faint-young-sun paradox—the notion that a warming sun hasn't correlated to an overheated Earth—is first discussed by C. Sagan and G. Mullen,

"Earth and Mars: Evolution of Atmospheres and Surface Temperatures," *Science* 177 (1972): 52–56.

A wealth of classic papers on carbon, climate, and atmosphere, including Arrhenius's from 1896, are republished and discussed in David Archer and Raymond Pierrehumbert, eds., *The Warming Papers* (Hoboken, N.J.: Wiley-Blackwell, 2011).

The famous BLaG paper is R. A. Berner, A. C. Lasaga, and R. M. Garrels, "The Carbonate-Silicate Geochemical Cycle and Its Effect on Atmospheric Carbon Dioxide over the Past 100 Million Years," *American Journal of Science* 283 (1983): 451–73. One seminal paper that preceded this (in science there are often many) is J. C. G. Walker, P. B. Hays, and J. F. Kasting, "A Negative Feedback Mechanism for the Long-Term Stabilization of Earth's Surface Temperature," *Journal of Geophysical Research* 86 (1981): 9776–82. More recently, an update of the model is in R. A. Berner and Z. Kothavala, "Geocarb III: A Revised Model of Atmospheric CO_2 over Phanerozoic Time," *American Journal of Science* 301 (2001): 182–204.

Maureen Raymo and her coauthors, W. F. Ruddiman and P. N. Froelich, launched a debate with the original publication of their uplift-climate hypothesis in M. E. Raymo, W. F. Ruddiman, and P. N. Froelich, "Influence of Late Cenozoic Mountain Building on Ocean Geochemical Cycles," *Geology* 16 (1988): 649–53; M. E. Raymo and W. F. Ruddiman, "Tectonic Forcing of Late Cenozoic Climate," *Nature* 359 (1992): 117–22; and M. E. Raymo, "The Himalayas, Organic Carbon Burial, and Climate in the Miocene," *Paleoceanography* 9 (1994): 399–404. This idea has very deep historical roots, deriving from some elements in T. C. Chamberlin's work of the late nineteenth century: T. C. Chamberlin, "An Attempt to Frame a Working Hypothesis of the Cause of Glacial Periods on an Atmospheric Basis," *Journal of Geology* 7 (1899): 545–84, 667–85, 751–87. See also Raymo's commentary in M. E. Raymo, "Geochemical Evidence Supporting T. C. Chamberlin's Theory of Glaciation," *Geology* 19 (1991): 344–47. For a general volume containing a number of different perspectives, see W. F. Ruddiman, ed., *Tectonic Uplift and Climate Change* (New York: Plenum Press, 1997). See also J. C. Zachos and L. R. Kump, "Carbon Cycle Feedbacks and the Initiation of Antarctic Glaciation in the Earliest Oligocene," *Global and Planetary Change* 47 (2005): 51–66, for references.

For a recent paper seeking to put the different lines of evidence together, see C. Garzione, "Surface Uplift of Tibet and Cenozoic Global Cooling," *Geology* 36 (2008): 1003–4. For a discussion of the geochemical issues related to the Raymo hypothesis, see S. E. McCauley and D. DePaolo, "The Marine $^{87}Sr/^{86}Sr$ and $d^{18}O$ Records, Himalayan Alkalinity Fluxes and Cenozoic Climate Models," in Ruddiman, *Tectonic Uplift and Climate Change*, 428–65.

A classic map of carbon dioxide levels over time is R. A. Berner, "Atmospheric Carbon Dioxide Levels over Phanerozoic Time," *Science* 249, no. 4975 (1990): 1382–86.

The interval prior to 45 million years ago was a hot one (known as the PETM, Paleocene-Eocene Thermal Maximum), and many have looked at the plants, carbon dioxide, and other factors at this time. An entrée to this literature includes F. A. McInerney and S. L. Wing, "The Paleocene-Eocene Thermal Maximum: A Perturbation of Carbon Cycle, Climate, and Biosphere with Implications for the Future," *Annual Review of Earth and Planetary Sciences* 39 (2011): 489–516; A. Sluija et al., "Subtropical Arctic Ocean Temperatures During the Palaeocene/Eocene Thermal Maximum," *Nature* 441 (2006): 610–13; J. C. Zachos et al., "A Transient Rise in Tropical Sea Surface Temperature During the Paleocene-Eocene Thermal Maximum," *Science* 302 (2003): 1151–54; J. P. Kennett and L. D. Stott, "Abrupt Deep-Sea Warming, Palaeoceanographic Changes, and Benthic Extinctions at the End of the Palaeocene," *Nature* 353 (1991): 225–29; and S. L. Wing et al., "Coordinated Sedimentary and Biotic Change During the Paleocene-Eocene Thermal Maximum in the Bighorn Basin, Wyoming, USA," in *Conference Programme and Abstracts: CBEP 2009, Climatic and Biotic Events of the Paleogene, 12–15 January 2009, Wellington, New Zealand*, ed. C. P. Strong, Erica M. Crouch, and C. J. Hollis (Lower Hutt, N.Z.: Institute of Geological and Nuclear Sciences, 2009), 156–62.

The paper describing the importance of new patterns of ocean circulation to Antarctica's climate is J. P. Kennett, "Cenozoic Evolution of Antarctic Glaciation, the Circum-Antarctic Ocean, and Their Impact on Global Paleoceanography," *Journal of Geophysical Research* 82 (1977): 3843–60. The timing of the freezing of Antarctica and its relationship to oceanic circulation are the subject of J. Anderson et al., "Progressive Cenozoic Cooling and the Demise of Antarctica's Last Refugium," *PNAS* 108 (2011): 11356–60.

Nate Dominy's papers on color vision and fruit include N. Dominy and P. W. Lucas, "Ecological Importance of Trichromatic Vision to Primates," *Nature* 410 (2001): 363–66; N. Dominy, "Fruits, Fingers, and Fermentation: The Sensory Cues Available to Foraging Primates," *Integrative and Comparative Biology* 44 (2004): 295–303; N. Dominy and P. W. Lucas, "Significance of Color, Calories, and Climate to the Visual Ecology of Catarrhines," *American Journal of Primatology* 62 (2004): 189–207.

The mobile field kit designed by Dominy and his colleagues is described in P. W. Lucas et al., "Field Kit to Characterize Physical, Chemical, and Spatial Aspects of Potential Primate Foods," *Folia Primatologica* 72, no. 1 (2001): 11–25.

NINE   COLD FACTS

For background on the military history of Project Iceworm and Camp Century, see E. D. Weiss, "Cold War Under the Ice: The Army's Bid for a

Long-Range Nuclear Role, 1959–1963," *Journal of Cold War Studies* 3, no. 3 (Fall 2001): 31–58.

The discovery of the causes for the ice ages, as well as the hidden climatic records in ice, have been the subject of fantastic general science books: John Imbrie and Katherine Palmer Imbrie, *Ice Ages: Solving the Mystery* (Cambridge, Mass.: Harvard University Press, 1986); Richard B. Alley, *The Two-Mile Time Machine: Ice Cores, Abrupt Climate Change, and Our Future* (Princeton, N.J.: Princeton University Press, 2002); and Doug Macdougall, *Frozen Earth: The Once and Future Story of Ice Ages* (Berkeley: University of California Press, 2006). All three are science writing at its best: authoritative, captivating, and well referenced. Detailed studies of ice cores, of the type described in Alley's *Two-Mile Time Machine,* reveal a complex set of cycles and oceanic events, each with their own names—Dansgaard-Oeschger cycles, Bond cycles, Heinrich events, and MacAyeal cycles. Climate can fluctuate wildly based on changes to glaciers, ocean circulation, and prevailing winds. Our understanding of the global extent and interworkings of the variables is a work in progress, aided by the increasing resolution available to geologists mapping chemical and physical changes to oceans and glaciers.

The impact of the ice ages on one part of human history is discussed in Brian Fagan, *The Little Ice Age: How Climate Made History, 1300–1850* (New York: Basic Books, 2001). For a beautiful account of how ice ages have affected both the landscape and life, see: E. C. Pielou, *After the Ice Age: The Return of Life to Glaciated North America* (Chicago, University of Chicago Press, 1991).

The work of Libby and Urey is discussed in Macdougall's superb *Nature's Clocks.*

An account of Dorothy Garrod can be found in P. J. Smith, "Dorothy Garrod as the First Woman Professor at Cambridge University," *Antiquity* 74 (2000): 131–36.

The importance of climate change and Natufian culture on the development of agriculture is the subject of debate, with the classical view in O. Bar-Yosef, "The Natufian Culture in the Levant, Threshold to the Origins of Agriculture," *Evolutionary Anthropology* 6, no. 5 (1998): 159–77; and O. Bar-Yosef and A. Belfer-Cohen, "The Origins of Sedentism and Farming Communities in the Levant," *Journal of World Prehistory* 3 (1989): 447–98. Other views—including contrary ones—are discussed in M. Balter, "The Tangled Roots of Agriculture," *Science* 327 (2010): 404–6.

The ways that diet, particularly the origin of agriculture, has influenced the structure of our genome are discussed in Spencer Wells, *Pandora's Seed: The Unforeseen Cost of Civilization* (New York: Random House, 2010). Jonathan Pritchard's seminal article on selection in the human genome is B. F. Voight, S. Kudaravalli, X. Wen, and J. K. Pritchard, "A Map of Recent Positive Selection in the Human Genome," *PLoS Biology* 4, no. 3 (2006). See also P. Sabeti et al., "Genome-wide Detection and Characterization of Positive Selection in Human Populations," *Nature* (2007): 913–88; and D. J. Wilson et al., "A Population Genetics-Phylogenetic Approach to Inferring Natural Selection in Coding Sequences," *PLoS Genetics* 7, no. 12 (2011).

TEN   MOTHERS OF INVENTION

The role of climate change in the origin and early evolution of humans and their closest relatives is evaluated in National Research Council and Committee on the Earth System Context for Hominin Evolution, *Understanding Climate's Influence on Human Evolution* (Washington, D.C.: National Academies Press, 2010), http://www.nap.edu/catalog.php?record_id=12825#toc. This volume contains an extensive set of references on the climate reconstruction. See also T. E. Cerling et al., "Woody Cover and Hominin Environments in the Past 6 Million Years," *Nature* 476 (2011): 51–56.

The hominid fossils from Chad are described in M. Brunet, "A New Hominid from the Upper Miocene of Chad, Central Africa," *Nature* 418 (2002): 145–51; and M. Brunet et al., "New Material of the Earliest Hominid from the Upper Miocene of Chad," *Nature* 434 (2005): 752–55. A recent analysis of bipedalism in early Kenyan finds is in B. Richmond et al., "*Orrorin tugenensis* Femoral Morphology and the Evolution of Hominin Bipedalism," *Science* 319, no. 5870 (2008): 1662–65. General books on the hominin fossil record include Ann Gibbons, *The First Human* (New York: Doubleday, 2006); and Donald C. Johanson and Kate Wong, *Lucy's Legacy: The Quest for Human Origins* (New York: Harmony Books, 2009).

Robert Merton's insights into invention are in R. K. Merton, "Singletons and Multiples in Scientific Discovery: A Chapter in the Sociology of Science," *Proceedings of the American Philosophical Society* 105, no. 5 (1961): 470–86; and R. K. Merton, "Priorities in Scientific Discovery: A Chapter in the Sociology of Science," *American Sociological Review* 22, no. 6 (1957): 635–59.

Stigler's law is in Stephen Stigler, "Stigler's Law of Eponymy," in *Science and Social Structure: A Festschrift for Robert K. Merton*, ed. Thomas F. Gieryn (New York: New York Academy of Sciences, 1980), 147–58.

For an account of how plants have influenced the history of life, see David Beerling, *The Emerald Planet: How Plants Changed Earth's History* (Oxford: Oxford University Press, 2007); and William C. Burger, *Flowers: How They Changed the World* (Amherst, N.Y.: Prometheus Books, 2006).

Steven Stearns's work on recent selection in humans is in S. C. Stearns et al., "Measuring Selection in Contemporary Human Populations," *Nature Reviews Genetics* 11 (2010): 611–22.

ACKNOWLEDGMENTS

My inspiration to enter science derived from growing up watching Apollo space missions on TV, visiting natural history museums, and reading writers such as Carl Sagan and Jacob Bronowski. And, as I grew, my parents, Seymour and Gloria Shubin, supported each hobby of the week—from rock collections and colonial pottery to telescopes and meteorites—no questions asked. They nurtured a child's curiosity, allowing it to transform into that of an adult scientist.

Kalliopi Monoyios, who created all the original art for the book, is everything you could want in a scientific illustrator: a student of the natural world who has a keen eye, sharp critical sense, and compelling aesthetic vision. Kapi helped transduce complex ideas and text into simple visuals. I had the privilege of having her in my lab for the past eleven years: she has now fledged to form her own freelance studio (www.kalliopimonoyios.com) and can be followed on her blog (blogs.scientificamerican.com/symbiartic).

I am fortunate to have worked in the field with remarkable people: Farish A. Jenkins Jr., Bill Amaral, Paul Olsen, Ted Daeschler, Jason Downs, Chuck Schaff. One of the joys of writing this book was reliving moments we shared in Greenland, Morocco, Canada, and Ellesmere Island.

Members of my lab past and present have been influential to and patient with my writing: Randy Dahn, Marcus Davis, Adam Franssen, Nadia Fröbisch, Andrew Gehrke, Andrew Gillis, Christian Kammerer, Justin Lemberg, Kapi Monoyios, Joyce Pieretti, Igor Schneider, Becky Shearman, Natalia Taft, and John Westlund.

For insights, comments, and responses to pesky queries, I thank Bill Amaral, James Bullock, Robin Canup, Sean Carroll, Michael Coates, Anna Di Renzio, John Flynn, David Gozal, Lance Grande, David Jablonski, Susan Kidwell, Andy Knoll, Michael LaBarbera, Dan Lieberman, Daniel Margoliash, Paul Olsen, Kevin Righter, Callum Ross, David Rowley, Paul Sereno, Michael Turner, Mark Webster, and Mike Young. Elena Skosey-Lalonde assisted with the fact-checking during summer breaks from the University of Chicago Laboratory Schools. Fred Ciesla patiently answered questions about the origin of planets many mornings after dropping our daughters off at kindergarden. Nate Dominy kindly shared stories and details of his work in Uganda. Lawrence Krauss graciously commented on the big bang and stellar formation, thereby saving me from embarrassing mistakes. Seymour Shubin, Michele Seidl, Kalliopi Monoyios, Andrew Gehrke, Joyce Pieretti, and John Westlund read and commented on drafts of the text. Thanks to all. The errors that remain are, of course, my own.

A number of the themes in the book were derived from interactions I had with students. Freshmen at the University of Montana, the University of Pennsylvania, and Skidmore College not only provided wonderful audiences but also offered questions that inspired me as I was writing this book. The same is true of students at a number of high schools, including those at Adlai E. Stevenson High School, Downers Grove North, the Francis W. Parker School, and the University of Chicago Laboratory Schools. I explored a number of the book's ideas in my own non-majors course at the University of Chicago.

My agents, Katinka Matson, John Brockman, Max Brockman, and Russell Weinberger, have been a continual source of support. Dan Frank and Marty Asher formed a powerful editorial team, helping me frame the book during rough drafts and polish it during later ones. Production, assembly, and copyediting derived from the often heroic efforts of Jill Verrillo and Ellen Feldman with the help of Ingrid Sterner, Teresa Gardstein, and Jenna Bagnini. The entire team at Pantheon has been a joy to work with.

My wife, Michele, kept things going at home during weekend writing escapes; tolerated many a date night discussing the Harvard Computers, Marie Tharp, and zircons; and read innumerable drafts of chapters, including those that ended on the literary equivalent of the cutting-room floor. Her endurance, intelligence, and love made this project possible. Michele and our children, Hannah and Nathaniel, are my deep connections to the universe I celebrate every day.

ILLUSTRATION CREDITS

Unless otherwise noted, all illustrations are by Kalliopi Monoyios.

8 Photographs of the Greenland crew courtesy of Bill Amaral; used with permission

20 Photograph of the "Harvard Computers," 1913, courtesy of the Harvard College Observatory; used with permission

37 Photograph of Beta Pictoris by the European Southern Observatory; rights granted under Creative Commons Attribution 3.0 Unported license (http://creativecommons.org/licenses/by/3.0)

47 Photograph from Zion Canyon on Earth by George Alexander Grant for the National Parks Service; image in the public domain. Photograph of Victoria Crater on Mars by NASA/JPL–Caltech/Cornell University; image in the public domain

72 Portrait of Seymour Benzer courtesy of the Archives, California Institute of Technology; used with permission

79 Photographs of Hindostan limestone tombstones courtesy of the Indiana Geological Society; used with permission

82 Portrait of Elso Barghoorn courtesy of the Harvard University Archives, Harvard University Press; used with permission

87 Portrait of Anton van Leeuwenhoek by Jan Verkolje (I); image in the public domain. Illustration of van Leeuwenhoek's microscope courtesy of Michael W. Davidson at Florida State University; used with permission

88 Portrait of Galileo Galilei by Domenico Cresti da Passignano; image in the public domain. Etchings by Galileo Galilei, 1638, in the public domain

93 Portrait of Preston Cloud courtesy of the Department of Special Collections, Davidson Library, University of California at Santa Barbara; used with permission

102 Photograph of *Glossopteris* fossilized leaf courtesy of the Swedish Museum of Natural History. Photographer: Yvonne Arremo; used with permission

103   Alfred Wegener portrait courtesy of the Alfred-Wegener-Institut, Germany; used with permission

105   Harry Hess portrait courtesy of the Department of Geosciences, Princeton University, Princeton, N.J.; used with permission

107   Photograph of Bruce Heezen and Marie Tharp from the Marie Tharp Estate, courtesy of Fiona Schiano-Yacopina; used with permission

108   Map by Bruce Heezen and Marie Tharp; painted by Heinrich Berann; used with permission

113   John Tuzo Wilson portrait courtesy of the Ontario Science Centre (www .ontariosciencecentre.ca); used with persission

122   Photographs of cliff in Nova Scotia by the author

123   Portrait of William Smith by Hugues Fourau; image in the public domain. Photograph of John Phillips, 1907; image in the public domain. Map by William Smith, published 1815; image in the public domain

142   Photograph of petrified forest reproduced with the permission of Natural Resources Canada 2011, courtesy of the Geological Survey of Canada (Photographer: Lyn Anglin). Photograph of fossil wood from Kaelin et al., "Comparison of Vitrified and Unvitrified Eocene Woody Tissues by TMAH Thermochemolysis—Implications for the Early Stages of the Formation of Vitrinite," *Geochemical Transactions* 7 (2006): 9. Image used under Creative Commons Attribution 2.0 Generic license (http:// creativecommons.org/licenses/by/2.0)

159   Camp Century photograph from A. Kovacs, "Camp Century Revisited: A Pictorial View—1969," *Cold Regions Research and Engineering Laboratory Special Report* 150 (July 1970): 44, 49; image in the public domain

161   Photograph of Louis Agassiz circa 1860; image in the public domain

164   Etching of James Croll by unknown artist, published in *Popular Science Monthly* 51 (August 1897): 445; image in the public domain

165   Portrait of Milutin Milankovitch by Paja Jovanović (1859–1957); image in the public domain

176   Photograph of Dorothy Garrod courtesy of Mrs. Kennedy Shaw and her daughter, Mrs. Caroline Burkitt. Photograph in possession of Pamela Jane Smith; used with permission

INDEX

Page numbers in *italics* refer to illustrations.